T0107367

# THE SHAPE OF A CITY

# THE SHAPE
# OF A CITY

*Julien Gracq*

TRANSLATED BY INGEBORG M. KOHN

TURTLE POINT PRESS  NEW YORK

LA FORME D'UNE VILLE

COPYRIGHT © 1985 LIBRAIRIE JOSÉ CORTI

ISBN 1-885586-39-6

LCCN 2004113838

DESIGN AND COMPOSITION BY JEFF CLARK AT

WILSTED & TAYLOR PUBLISHING SERVICES

# THE SHAPE OF A CITY,

as we all know, changes more rapidly than the heart of a mortal. However, it often happens that before being discarded, left behind to become the prey of its memories, the city—caught, like all other cities, in the vertiginous metamorphosis that characterizes the second half of our century—will have found ways to change a heart still young and impressionable just by subjecting it to its climate and landscape, and by leaving an imprint of its streets, boulevards, and parks on the most private thoughts and daydreams of its owner. It is not necessary to have lived there like an ordinary citizen; I even doubt that it would make much of a difference. The city's influence will be much stronger, and perhaps last longer, if it has remained partially hidden—if, because of some unusual circumstances, we have lived in its midst but never reached a degree of familiarity, much less of intimacy, if we never had the freedom, nor enough leisure time to walk through its neighborhoods aimlessly, to stroll its streets at will. It is possible that by making only certain concessions and without ever completely surrendering, the city has—just like a woman—tightened the

1

threads spun by our daydreams around herself, and better adapted the rise and development of our desires to her rhythms and moods.

To live in a city means weaving one's daily peregrinations into a maze of paths usually linked around several directional axes. If one disregards all movements connected with one's job and counts only those steps leading from the center to the periphery and back again, it becomes clear that Ariadne's thread, which ideally unravels behind the true city dweller, takes on the characteristics of a carelessly wound skein of wool. It encloses an entire complex of streets and squares within a finely meshed network of comings and goings; seldom do we wander into outlying areas, venture forth beyond the familiar haunts. There is not the slightest similarity between the plan of a city and our mental image of it as we consult the unfolded map, or between the sediment deposited in memory by our daily wanderings and the sound of its name. The Paris where I lived while a student, and later on during my mature years, is contained in a rectangle bordered in the north by the Seine, and almost entirely by the boulevard Montparnasse along the south. This heart of an area which comes alive during my daily peregrinations is surrounded by concentric circles where—at least in my perception—all activity progressively decreases until, close to the periphery, they become lifeless, indistinguishable from each other. It is the central chambers of a labyrinth that attract the city dweller like a magnet, a locale he returns to time after time; its perimeter serves mostly as a protective screen, an insulating layer intended to shield the inhabited cocoon, and to prevent any osmosis between the outlying areas and the true city life securely locked within that inner sanctum.

This is not how I lived in Nantes. The regime of the boarding school, during the nineteen-twenties, was strict. We were forbidden to leave the premises except for vacations, and on Sundays; furthermore, it was required that *a designated person* come to collect us in the parlor, and also deliver us there in the evening. I left the school only once every fortnight; the rest of the time, all I could see of the city were the crowns of the magnolia trees in the Botanical Gardens beyond the walled-in courtyard and, whenever the gates were opened to let in the day students, at five minutes before eight in the morning, and at five minutes before two in the afternoon, a brief glimpse of the museum's façade. But this strict reclusion was like a one-way street. Twice a day, like the tide, the sounds of Nantes poured in together with the wave of day students, sometimes just filtering through, sometimes in a loud blast. I lived in the heart of a city that loomed large in my imagination, but which I did not know very well. I was aware of certain landmarks, and familiar with some itineraries, but its substance, and even its smells, never lost their exotic flavor; a city where all the views led only to ill-defined, unexplored, faraway vistas, a loose framework easily absorbed into fiction. Each of the rhumbs of this compass shone brightly but separately in my imagination, like lone stars not part of any constellation.

I was eleven years old, and just starting sixth grade. Half-known, half-imagined in my dreams, the city has never cast off that impression I've kept since our first contact. Later on, I lived an entire year in Nantes under ordinary circumstances, in order to complete my military service as a second lieutenant, and then returned again to teach in the same lycée I had attended as a student. How odd that not a single distinct memory has survived to remind me of the new familiarity gained during that time, which evaporated as soon as it was interrupted. Today, I would get lost in the streets of the neighborhood where I lived during that year, I could not even recognize the house where I lived; it seems as if the time I lived in Nantes *like any-*

*one else* never existed. Something that at first haunts you with all the charms of a faraway, unattainable princess will evaporate later on during the disenchantment of cohabitation.

A city which remains for such a long time almost "off limits" eventually becomes the very symbol of freedom. Every time I am back wandering through Nantes, the brisk breeze still blowing through its streets refreshes them pleasantly and for me alone: the city, anchored in solid ground that supports it securely on both sides of its strangled estuary, is already subject to an inland, rather than seaside, climate, quite different from that of Quimper, where I lived later on, whose coastal sky changes with the tidal flow. It is only in La Fosse, a residential area located almost on the outskirts of Nantes, that one senses the presence of the sea; here is where the riverbed suddenly deepens into a channel, signaling the end of a stream which up to this point flowed leisurely through the countryside, holding its own against encroaching green meadows, and which, in my youth, still lingered around the islands of Nantes just like a *varenne tourangelle* (one of those plains near Tours inundated periodically by the Loire). It was not the sea breeze that aerated the streets, but that mental lightheartedness which takes hold of us at every intersection where, in our imagination, the unpredictable lies in ambush. At the end of childhood, at the end of adolescence, powerful images infiltrate the mind, goals deemed attainable beckon and vie for our attention so vehemently they trigger dizzy spells, leaving us overwhelmed by the enormity of what is left unaccomplished at the end of each day, by the prospect of so many *projects abandoned or rejected*. Life just outside the school gates, out of reach and yet so close I could feel its heartbeat, was passing me by, leaving me stranded; it animated the streets of the city, but all I could do was listen obsessively to its sounds: it is the memory of those noises, so close, electrifying, and nevertheless impossible to reach, which remind me more than anything of certain poems by Rimbaud, such as *Workers* ( *"The city filled with smoke and the noise of workers plying their trades followed us very far on our paths ..."*), poems that express those cold shivers felt when hitting rock bottom, that passive stupor of suburbs living with their ears pressed against the muffled beating of a heart. On certain evenings at the start of summer, when the heavy, sugary, vegetal odors of the Botanical Gardens came wafting across the street, the proximity of that concentration of activities inaccessible to us made our heads spin. Patches of yellow light cast by

6

the setting sun still lit up parts of the dormitory where we took our clothes off: a feeling that the day was closing in on us too soon, that the streets, oblivious of the curfew imposed on us, were just awakening, and that unusual activities, more mysterious than work as we knew it, were about to begin kept sleep at bay for a long time in the double row of our iron bedsteads.

It is not my intention to draw the portrait of a city here. I would simply like to show—with all the clumsiness, inaccuracy, and fiction inherent in such a return to the past—how Nantes has shaped me, that is to say, both prompted and forced me to look at the world through a prism which distorted my vision, a world just taking form in my imagination thanks to the books I read; and how, since my seclusion afforded me greater freedom to make use of its landmarks, I have reshaped it to fit the contours of my secret dreams, breathed life into it according to the laws of desire rather than those of objectivity. May it thus accompany me, like some *vade-mecum* one carries around everywhere to carelessly leaf through, annotate, and erase, a familiar repertory always available for quick consultations, a springboard unsuitable for plunging into fiction but also a network of mental ruts carved and hardened inside of me, the result of wanderings and explorations the city compelled me to undertake. It just so happens that the course of events during the last fifty years protects me against any argument based on facts that run counter to the image I carry within me of that city, the incubating milieu of my adolescence. Like in many other French cities —and here to an even greater degree—the bombings of 1943–44 have reshaped its physiognomy. But even more remarkable was the metamorphosis that took place when I graduated from the lycée. The eddies of the Loire surrounding the islands were being filled in, as well as the riverbed of the Erdre which once flowed downtown, changing the city's foundation forever. At the same time, underground tunnels were being dug for the trains which used to run through the heart of the city, passing through three consecutive railroad stations just like in a *miners' settlement* of the Far West. At the very moment when, at age eighteen, a period of my life came to an end, the city closed off and sealed an entire block of my memories with these major municipal projects. It is strange that instead of causing a profound sense of separation, this solution to a problem involving the continuity of a place and of a life eliminates from my ruminations about Nantes the touch of worry when trying to recall what lies buried in the past. But, perhaps, not so strange after all ... The acrimony that accompanies ruminations about aging arises from the fact that our memories are situated in settings which have remained intact: it is the immutable youthfulness of the world that makes the underlying decay and decrepitude it supports difficult for us to take in stride. Nothing of that sort happens when I find myself walking again in Nantes. My old town—my old life—and their new versions do not succeed each other in chronological order but overlap in my mind. The energy flowing between them transcends time, memory is set free of melancholy and gravity's weight; the feeling of a reference outside linear time propels me forward, fusing images of the past into the present rather than pulling my mind backward. I would like that nostalgia, which I am tempted to indulge in like anyone else, be absent from these pages. Sheer chance has made the years of my childhood and adolescence into a deposit of great value, riches at my disposal that I spend freely, without ever feeling any poorer. So let us return and wander the streets of Nantes, not in order to indulge in a sentimental journey down memory lane, but to find out who I became because of them, and vice versa.

# I CANNOT SAY WHY NANTES

has always been my town without first stating the reasons why Angers is not, and never has been. In spite of the fact that everything—the conveniences, the distance, the administrative services, family and business relationships—linked the Saint-Florent of my childhood to Angers, the official regional capital, where my father and mother went to secondary school, as did all the children of the noteworthy, or semi-noteworthy Florentines. The route, the railroad stations along the line leading to Angers, interrupted midway by a stop in Champtocé (birthplace of my mother and home base of the entire Belliard clan) became familiar to me at an early age: like a yardstick for measuring fabric, I still use them to calculate traveling distances. One of my most vivid memories dating back to the First World War is the local night train serving that route: smell of wet coal, hot-water bottles, pale light cast by small, yellowish lamps, mirrors forever trembling inside their shaking wooden frames while the rain outside cuts swaths through the darkness, stops during nights drowsy with sleep where a voice repeatedly calls out the name of some station impossible to understand before fading into the fog. What I always enjoyed—and still enjoy now, whenever I travel there in the daytime—is the trail of residential suburbs, stretched out along the banks of the Loire, from la Possonnière to Béhuard, and from Béhuard to Bouchemaine, three stations signaling the approach of the town. Scattered over a tapestry of gently sloping vineyards, goat paths that climb steep hills between the remnants of stone walls, stumpy blocks of limestone and gray slate, patches of scrub brush burned by the summer sun, there beckons an array of country houses ranging from the Turkish-Hindu pavilion to the imitation Trianon villa, buildings set in the hillsides overlooking the Loire reminiscent of the half-baroque, half-fantastic architectural style of seaside resorts popular during the Belle Epoque. Even today, I love to see that parade of summer homes along the railroad track—the sharp angles of their little towers and gables jutting into the bright yellow summer sun just setting on the horizon—a succession of *follys* spread out next to the river like the long, shimmering train of a gown, indicative of a city's importance and creativity: two promises that town has never kept. I always thought that the genius of Angers—if there ever was a genius—must be confinement: the choice of its location, a stingy, unattractive site on the banks of a

**11**

minor tributary away from the great river, brings to mind those narrow-minded souls who are afraid to build a settlement in the midst of open spaces, and crowd their houses and churches into an inconspicuous corner of the planner's drawing board. All of the city's alliances and exchanges—a provincial capital late to become industrialized—are of limited scope; throughout my childhood, they never went beyond the regional level. There was the local gentry who spent the winter in their town houses in the rue des Arènes and then returned to their manors in the farmlands criss-crossed by hedges and trees, villagers— with long lists of *errands to run*—on shopping trips, notaries and other country folk in town to take care of legal matters, black-robed priests and wimpled nuns of a province rich in teaching, charitable, and other religious orders making their rounds of *ad limina* visits. A stranger wandering the streets of Angers is likely to sense its shortcomings almost immediately, especially if he ventures into the deserted neighborhood around the cathedral and walks through its narrow, empty streets with their sleeping cats and potted geraniums where, in years past, one might catch a glimpse of a priest's robe silently swaying in the distance. I always thought that Angers, with its well-hidden intrigues, its unmistakable odor of social *withdrawal*,

12

its stagnating, miserly, bickering micro-societies, was the true homeland of *Le Curé de Tours*, rather than the city chosen by Balzac. Certainly more Balzacian than Nantes, when presented in this light: Balzac is the novelist who best analyzes pockets of social stagnation—with the exception of those he found in Paris, scandalous in his eyes because he thought of that city as a cancerous *implant* in the weak vascular fabric of rural France—places where the air is stifling, where the physical environment and the lack of oxygen become so oppressive they generate waves of debilitating radiation, like those released by the house of Usher on its last owners. Maurice Fourré's entertaining novels remind me of the particular way people settle and put down roots in Angers; he knew and celebrated Angers as well as Nantes, but his heart belongs to the former. Whether his fiction is set in Paris, Richelieu, or Brittany, I always detect the playful, talkative *farniente* of the Angers bourgeois who divides his time between the shop in the rue des Lices, the little farm house built of *tuffeau* stone, and the vineyard with its summer storage cave on one of the Loire's hillsides, who goes fishing on Sundays or gathers in some leafy bower with his fellow *boules de fort* players, a special game of odd-shaped balls men played barefoot around the turn of the century. Hardly overburdened as an administrative center, home to more notaries than businessmen, a discrete digestive organ of income derived from real estate holdings—but also neat and clean, flower-bedecked, inviting, its heartbeat slightly lowered, as if local working days had been infused with, lightened by supplementary leisure time—the city on the banks of the Maine has taken measures to secure for itself a more than comfortable retirement life, and to avoid American-type *stress*. It seems to me that even as a child, I was already dimly aware that revenue collected from tenant farmers was accumulating right here in the city where it circulated still in its raw form, undisguised, not yet converted or invested like elsewhere, accentuating in my mind the image of a clumsy town with mud still sticking to its shoes, a town not yet completely segregated from its farmlands.

The city has changed since my childhood days. It has become more lively; the old, almost peasant-like nonchalance has been bartered for a businesslike ambiance that lacks a solid infrastructure, so that many a walk around town turns out to be a disappointment. Nevertheless, I like to spend the time between trains (Angers has been for a long time, and still is, as far as I am concerned, a city "between two trains"—ever since that wintry day when, sitting on a bench, I had cut open the pages of *The Cliffs of Marble*) at the castle, walking on the right side of the moat all the way up to the mall squeezed into a narrow space that comes to a dead end, high above the Maine municipal gardens laid out down below along the riverbank where once stood a row of ancient, ugly houses. The castle, firmly planted on its spot like the solid mold of a child's sand pail turned upside down, is for me—together with the cathedral of Albi—the most beautiful, massive work of masonry I have ever seen in France; the dark gray stone set in cement has a pleasing effect on the eye, a combination of materials also used successfully in the construction of the snack bar at the new railroad station. Looking straight down into the castle's shadowy moat, carpeted in a shade of green reminiscent of pre-Raphaelite paintings, offers a charming, almost magical surprise—the sight of a herd of spotted deer peacefully grazing. I am saving the exploration of the rue de Létenduère for a day of leisurely walks; its gentle slope from the train station down to the housing developments along the Loire invites and intrigues me: I would like to see how, at this late date, the city has finally joined the stream which had intimidated it for such a long time. Listening to comments by first-time visitors who find Nantes repulsive and Angers enchanting, I sometimes have the impression of being unfair. But there is this persistent feeling, stronger than any other, that nothing will ever attract me to Angers: an attitude just as cutting and unjust as the indifference shown by a man who decides, after just a momentary glance at a woman, that nothing about her will ever appeal to him.

Moreover, since the adult's conception of the world is indebted to ideas formed during childhood, there are two other reasons why I never could, and still cannot, consider Angers a full-fledged city. The first concerns statistics. Since I lived in a very small provincial town, my viewpoint was essentially bookish, and based on numbers. In those times, when elitism had not yet become the object of an efficient counteroffensive, manuals ranked French cities the same way our seating assignments were decided at school, that is, according to our compositions' order of excellence as reflected in the number of points obtained. Or, I then was of the opinion that a genuine city, *a city teeming with life, the city of one's dreams*, did not reach a level of excellence comparable to that of a perfectly executed writing assignment (then graded as brutally as an exam) unless it went beyond a certain threshold where quantity abruptly transformed itself into quality, something that became fixed in my mind once and for all by a six-figure digit: one hundred thousand inhabitants. At that time, a census had left Angers languishing ingloriously at around eighty thousand: it shared, and still shares, to a lesser degree—together with Amiens, Montpellier, Besançon, and Grenoble—the disgrace afflicting those students whose quarterly report cards indicate cause for concern by reminding them that they *need to improve*, and who later on, as civil servants, one sees stagnate eternally in subordinate positions.

The other reason concerned the question of *tramways*, then an item of my personal mythology.

Until I was fifteen, and even beyond that age, one of my bedside books—together with the periodical *Le Chasseur français*, whose descriptions of itineraries for bicycle tourists I devoured—was an old Michelin Guide found in the attic next to a collection of the Vermot Almanac, sustenance for more than one afternoon of enforced fasting when I had run out of Jules Verne, as well as Fenimore Cooper. Even today, when I reopen the considerably heftier Michelin Guide, its size having steadily increased over the years so that now it looks like a small telephone book, I can get carried away for an hour or two just looking, fascinated, comparing and superimposing in my imagination the tracings of those pink city maps which long ago were just about the only halfway accurate images I had of France. But then I only had eyes for a few of them: if the plan did not show the dotted lines illustrating a network of electric tramways, I lost all interest and turned the page; there was nothing to do but *move on.*

It is difficult for me to say how this bizarre prejudice had come about. Unaware of the rationale of transport systems, I probably thought that a city without tramways was the equivalent of a country without trains: not a consequence of size, but an indication of a civilization's staggering lack of progress. I am not sure, but there might have been other reasons: the prestige of miniatures like model trains, which in my opinion also included the local railway system, whose placid, nonchalant cars parading by looked like intermediaries between a real train and a toy—as well as the impression of riding on a finely tuned merry-go-round that turned on its tracks in a limited space, a conveyance perhaps more entertaining than utilitarian, so that every time I boarded a tramway I felt as if on holiday, free to enjoy a day off work. I must admit that several cities with less than one hundred thousand inhabitants found grace in my eyes because of their tramways. But not Angers. Small, skimpy, high up on their wheels, serving only a limited number of routes, the angevine tramways never impressed me: those of Nantes were longer, more streamlined, of a pleasing, buttery color, they cleared the road with an arrogant, authoritarian clang, inviting an immediate comparison between a sleek, rapid locomotive and those rumbling old battered engines about to finish their career by pushing railroad cars on the tracks of a switchyard. Moreover, at the onset of summer, visitors were pleasantly surprised to see a completely open trolley called a *baladeuse* hooked up behind the engine, a wagon with neither windows nor partitions between the seats and without the customary entry and exit spaces, sitting low on the tracks, immediately accessible right from the road, and as easy to board and leave as an escalator. These charming vehicles, wide open to the wind and the sun, announced summer in the streets of Nantes as ritually as the cuckoo announces springtime. Although they disappeared more than a quarter of a century ago, their memory has kept my secret attraction to Boris Vian's novel *Autumn in Peking* alive, where a tramway full of passengers has escaped from its dull urban route like a planet gone astray, and continues traveling up and down the coun-

19

tryside without ever coming to a stop. For my part, I would have gladly gone to explore the highways and byways of France aboard one of Nantes' seductive *baladeuses*; every ride was an adventure, just as if I were playing hooky. Isn't that what happened also to travelers in one of my favorite novels by Jules Verne, aboard *The House of Steam* which roamed the roads of India?

However, while I am writing these lines tinged with regret, a local newspaper publishes surprising news: the tramways of Nantes are coming back! The last city to have done away with them, but enlightened since by who knows what oracular council, will be reactivating its lost talisman.... News quite appropriate to confirm my idea, ever present while I write these pages, that time is reversible, that it is possible to resurrect Nantes' past, where the unequal paving stones which once covered its streets do not guarantee the rise of *the immense edifice of memory*, but where those years of exalted anticipation still maintain a dialogue with present and future times. That certain period of my past, those seven years spent mostly dreaming rather than living, is only half asleep: whatever remained unaccomplished during a life half-cloistered continues its subterranean burrowing deep inside of me, like rhizomes which from time to time unexpectedly grow a shoot that thrusts upward and breaks through the earth in the form of a green stalk.

At first, and for a long time thereafter, Nantes was only a simple stopover on our way to the seashore where we spent our summer vacation. The train, which then crossed the heart of the city at the speed of a river barge, ran alongside one of the Loire's eddies, and stopped at the stations Nantes-Orléans, Bourse, and Chantenay. It slowed down traffic; but the curious traveler, drawn to the train compartment's window by the din of the streets and the dockside, was rewarded by getting an unusually close look. Instead of offering the usual views of empty lots, scrap metal warehouses, inner courtyards of buildings with their garbage cans and gardening tools, the train suddenly split the inner city right down the middle before the surprised traveler, like an anthill bisected by the blow of a spade, amidst teeming traffic that coagulated along the railroad tracks into instantaneous clots at every train crossing. Twenty years later, this long-forgotten childhood impression surfaced again quite unexpectedly in one of Lille's suburbs—it could have been Menin—where the train taking us back from Holland had stopped for a moment, at the end of May 1940, in the midst of the great collapse. Refugees on their way south had gathered outside the railroad crossing; on the opposite side of the track, northbound refugees were on their way home after fleeing the Somme, where the Germans had already begun to cut off the passage. I was overcome by that same feeling of mounting agitation, of activity gone out of control I had experienced as a child while crossing Nantes; it brought into focus, rendered specific the shade of malaise and dizziness that had colored my first contact with the big city. More than once, an inner turmoil, at first incomprehensible, is a sign that a certain encounter will be decisive for us; but the compass needle, panic-stricken, will keep on spinning for quite a while before pointing to the metallic mass that has upset it.

The movement of this convoy as it progressed slowly, lazily through the very center of a big city to the ringing of bells at railroad crossings, the hurried clanging of tramways, and the concert of horns and whistles awakened in me the feeling that here, for the first time, I had come face to face with life lived on a large scale, a life full of violence, haste, and devilish jubilation. For a child growing up in the country, the city's big surprise is not so much its physical impact, the unexpected scale of buildings and streets, and the

22

profusion of unusual objects, but rather the over-whelming, completely new feeling of being suddenly submerged in a sea of seething humanity, something the slow-beating pulse of Angers had been incapable of communicating to me. A child growing up in the city is impervious to a moment like this when life goes to your head like strong wine, a sudden impact just as decisive, and as bewildering in its effect, as the onset of puberty. Nature's protective, gently cradling and swaying rhythms that carry you along come to an abrupt halt when interrupted by an unexpected con-frontation of *recklessness*, a premonition of the hu-man jungle. Even if I tried, I know that I could never liberate myself from this ambivalence Nantes awak-ened inside of me, and which memories of Menin have since reinforced: whenever I find myself in the midst of unruly crowds, I am still the child clinging to the train compartment's window, speechless and unable to move while watching, from up close, the violent ag-itation of a city cut in half like a worm.

We would stop in Nantes only long enough to change trains; but on the way back from our summer vacation, my parents would sometimes take advan-tage of the stopover and spend a few hours to place an order, or to visit a supplier. To this I owe the delight-ful memories of my very first tramway ride (whereas my first encounter with the railroad left a blank): first to Chantenay, to the offices of a soot-encrusted factory, and then a second ride to Pirmil, where my parents placed an order at the *Bertin* soap factory. I can still see the label that decorated one of its products: worthy, because of its name and wrapping, of the perfume shop in *César Birotteau*, and without doubt commemorative of the French victory at Agadir, since it was called *Soap of the Princes of the Congo*. We went on a guided tour of the soap factory, a building four or five stories high, whose windows looked out over one of the arms of the Loire; the noise, the commotion of the conveyor belts, the clean and well-lit assembly and packaging rooms, and the penetrating, slightly sweet odor of fresh laundry define my first contact with the world of industry: a contact neither repulsive nor depressing. I've never had the opportunity to get a closer look at mechanical, assembly-line work, and one may very well think that I am speaking of it lightly; but in fact there is nothing *Dickens-like* in the spontaneous image which comes to my mind, and which tends to join directly those bright and glossily painted factories of the nuclear age set on carpet-like green lawns—taking a giant stride over the leprous era of coal dust and soot which was still very much a reality in my youth. All this happened during the war years of 1914–18; these images of the tramway, the soap factory, the glorious, majestic procession of the train alongside streets where nothing seemed amiss except a row of cheering spectators are my first memories of Nantes. If the picture darkens from time to time, it is because of shadows cast by so many tall buildings, or the cavernous aspect of streets, which surprised me. All things considered, what lingers on after that first, fleeting contact is—rising from its resonant, shaded, freshly washed streets teeming with life and laughter, from the crowded terraces of cafés in the summertime, refreshed by the misty scent of lemons, strawberries, and grenadine, smells inhaled while walking about in that city where life's heartbeat was no longer the same and which has, since then, remained unforgettable—an unusual, daring perfume of modernity. For me, that perfume has been, and always will be, associated with a season, my favorite season, a time when all the secret, almost erotic powers of the city are released. Later on, I certainly have loved Nantes bundled up and locked inside heavy winter fogs, a time when vendors of roasted chestnuts and black wheat flatbread set up their stands at street corners, every hole of their perforated roasting pans glowing red-hot from the fire within. But summertime will always be for me, ever since my

25

first contact with her, the fateful season of the city called Nantes *la Grise*. As soon as the pink and white candles of the chestnut trees start to light up the avenues, as soon as the leaves of the magnolia trees in the Botanical Gardens regain their brilliant sheen, these hardly noticeable signs of my favorite season go right to my head, setting in motion something which even a fully orchestrated explosion of springtime in the country could not make me feel: I sense a sudden softness in the air, the sensual warmth rising from a rumpled bed, flowing through the streets, for me alone.

# A BOULEVARD WITH NO

traffic, no businesses": this annotation referring to a poem in one of my favorite books, the *Illuminations*, brings back memories of a neighborhood in Nantes without a single large boulevard but nevertheless redolent of a kind of drowsiness exuded by wealth, of siestas taken amidst well-tended flower beds, and of summertime yawns discreetly stifled in the shade of stately residences located in affluent neighborhoods (Rimbaud's poem and annotation date from Brussels: *Bruxelles—Boulevard du Regent*). The lasting image of a city where one has lived for a while tends to grow—in a prolific, anarchical manner—from a single, germinal cell which does not necessarily coincide with a functional, or nerve, "center." This cell, this limited area from which everything fanned out and returned as rigidly as arrival and departure times set at boarding school, was the administrative, military, and clerical center of Nantes, whose north-south axis, between the river Erdre and the Loire, follows the avenues known as *Cours*: cours Saint-Pierre and cours Saint-André. Its center is a monument hardly less exotic for France than the Obelisk, the highly irrelevant and very solitary *Colonne Louis XVI*; all around it, inside a circle of approximately three hundred meters in diameter, stood the governor's office building, City Hall, the cathedral, the museum, the lycée, the local army headquarters, the Botanical Gardens, and the Château. It is a quiet, porous neighborhood with little traffic, where life seems to burrow into the ground and then surfaces again, bubbling up toward the periphery; the vital current coming from the rue du Maréchal Joffre pauses here for a brief rest at the foot of the one and only French monument dedicated to the locksmith king, and then disappears into the rue de Verdun. How strange that after leaving one lycée to attend another (from Nantes' lycée Clemenceau to the lycée Henri IV in Paris), and then while at the Ecole Normale, I not only rediscovered in the neighborhood around the Panthéon an almost identical church-like silence, rituals practiced and regulations observed by people coming and going like clockwork, their cautious, measured movements reminiscent of those I had seen in Spain during the hours of the siesta, but was once again overcome by the very strong feeling that this was a public area designed on too large a scale, too vast for the anemic trickle of life that never quite succeeded in its efforts to animate it, a feeling first inspired by that part of Nantes which surrounded our school (a complex closely resembling a military compound). Today, when I walk from the train station through Cathedral Square toward the Bouffay *pedestrian zone*, I am again crossing an inner city area that has hardly changed—slightly morose, a little haughty, sparsely populated by small groups of people—the very image of a certain urban profile that imposed itself on me at age eleven. This could be just my personal impression; to a stranger who has also just arrived at the nearby train station and then crosses this neighborhood of *Cours*, it must look like a boring replica of Bordeaux's allées de Tourny, lost in a city that remains largely plebeian. Since childhood is known to spontaneously develop concrete images *beyond their immediate significance*, images which our affective memory then registers once and for all, I could already envision Delvaux's and even De Chirico's cities behind the Corinthian pilasters and the moldy pediments of those old townhouses slumbering at a safe distance behind their forbidding alignment of chestnut trees, lost in dreams about architectural feats no longer mysterious. Defying all odds, they are here to stay; I always find them again. I have not the slightest aversion to resonant, teeming, noisy cities excited by their own hustle and bustle, which sometimes hint at what might be an orgasm of pure, uninterrupted activity. This was my impression of Nantes on my first visit; I also remember, while traveling in Portugal, how thrilled I had been by that inexplicable *allegria*, an elation caused by the feverish liveliness of traffic in the streets of Porto. However, the ideal city taking shape spontaneously in my daydreams is the Nantes of its *Cours*, those long, tree-shaded avenues, a city evacuated until nothing remains to soften the sharp angles of its streets, where all human breath masking its stony arrogance has evaporated.

This atmosphere of distinctive, but rather lonely Sundays on the streets surrounding the lycée did not dissolve until one reached the neighborhood of the rue de Richebourg, then a winding and rather seedy, unsavory road on a lower level alongside the south façade of the lycée. In the summertime, the racket made by a lathe—probably a woodworker's—carried across that narrow street and as far as the open windows of the sixth-grade classroom; a background noise that will always remain in my memory as the obligatory accompaniment to our recitations of the litany of Latin declensions, as well as a disheartening sound effect lending additional meaning to the expression "être au rouet" (chained to the wheel). At the start of school vacations, while walking down the rue Stanislas Baudry to reach the train station, I would glance sideways in the direction of that depressing, crudely paved road, smelly and without sidewalks, very much like a street during medieval times. The lycée's chapel towered over it from up high, like the elaborate staterooms of a three-mast schooner seen from down below: even today, the confrontation of that haughty architecture built under the auspices of the Third Republic with a street of ill repute is a brutal reminder of the separation between the city proper of Nantes, where I lived in the newest, most anonymous neighborhood, and its southern border.

As a matter of fact, that neighborhood has hardly changed. The rue Richebourg has been cleaned up and sanitized, and proudly exhibits the insignia of a three-star hotel nowadays. However, a line like the hardened scar of a military front continues to run obstinately between the two sides of the street, one of those significant signs of demarcation that fragment a city into a maze of interwoven blocks, capable of affecting one's imagination just as strongly as the contrasts between the little colored stones of a mosaic. When, walking south, I passed the entrance to the rue de Richebourg and, on the opposite side, that of the rue Ecorchard, I was in effect crossing a border: behind me, there was the deep, richly oxygenated respiration of a French architectural garden, its pores largely dilated

3 1

in the luxury of space generously measured; in front of me, the agitation of a wan city toiling away in crowded quarters, whose prospects always seemed to get lost in a gray, noise-filled haze of horse-drawn wooden carts and hackneys at the end of boulevard Doulon. In spite of a tobacco factory's presence, a small enclave of elaborate industrial architecture not entirely unworthy of Ledoux, surrounded by streets named rue de Manille, rue du Maryland, and rue de La Havane (Nantes is fond of streets grouped by families), I have always considered the boulevard Doulon, now rebaptized boulevard Stalingrad, as well as all those avenues scarred by tracks fanning out or leading to the switchyard of a central railroad station, one of the city's repulsive zones. It seems to me that the rust, grit, and grime of abandoned railway tracks cast their mournful, gloomy shadows as far as on the greenery of neighboring streets, so different from the proximity of a cemetery which I find rather soothing, comparable to those areas designated as *zones of silence* in public parks.

Apollinaire was the first poet to notice that a sudden clearing in the weather creates a microclimate that can cleanse and illuminate a nondescript street, create an unexpected flash of happiness just by showing how a façade of houses can capture the new day's sun at ten o'clock in the morning:

I've seen a pretty street this morning whose name
    I forgot
Heralding the sun in its new, and proper attire

Although Apollinaire situates his street close to
the Ternes city gate ("I love the gracefulness of that
industrial street—located between the rue Aumont-
Thiéville and the Avenue des Ternes"), I always tend
to imagine that during his lifetime streets awakened
by such fresh, glorious rays of sunshine must have
existed primarily at the southwestern edges of Paris
—in the neighborhood of *Port-Aviation*, or around
Issy-les-Moulineaux—where, around 1910, industrial
buildings in the form of movable hangars had sprung
up to house internal combustion engines and the first
airplanes, shelters just as light as those birds made of
wood and sailcloth which they protected, fueled, and
repaired. Even today, standing on the corner of certain
streets in Boulogne or Billancourt which overlook the
Seine, I am surprised to find in their perspectives an
air of unexpected neatness, as if freshly swept clean,
which the slightest touch of the morning sun exalts
and almost sets aglow; one might say that the naïve
enthusiasm of workers belonging to the *heavier than
air* generation still lingers here. There is always that
element of surprise when, while walking down streets

33

one expects to be ugly, marred and disfigured by the most degrading forms of manual labor, we suddenly see them transfigured by a ray of sunshine—like a moment of fleeting happiness. But such a surprise caused by the most insignificant event or impression can also happen elsewhere, in the most banal bourgeois neighborhoods, although there it is not as easily explained: it could be an unexpected declivity in the road which invites, tempts one to continue in that direction, a very slight turn of a road's axis which both veils and partially reveals a perspective, a tree leaning over the sidewalk from above the crest of an ancient wall, a pleasing harmony in the rhythm of buildings alternating with free spaces which suddenly catches the eye. Instances when we are overcome by a feeling of how wonderful it would be to linger there, assured that life has regained its normal pace and recovered its guideposts, and that the universe has found a way to renew us and confirm its promises with just one brief, smiling look.

Not long ago I thought of these fleeting moments of happiness in urban settings while taking a stroll up the rue de Charost in Ancenis, a road where I like to walk during late morning hours in the direction of the fairgrounds; it is like the secret garden of stones in a town which, ever since childhood, has been the very symbol of provincial boredom for me. I always think of it when I find myself again on the rue Clemenceau in Nantes, because every time I stepped out on the sidewalk after closing the door of the lycée behind me, I could only think of it as the fateful abode one leaves vowing never to return; and yet, at the same moment, I would see the tropical trees of the Botanical Gardens down the street project the outline of their crowns against the sky, an exotic sky filled with the wondrous forebodings illustrated by Jules Verne.

Nantes' Botanical Gardens—with the sole exception of a small monument to Jules Verne featuring a bas-relief showing the moon, a balloon, a volcano, and a viaduct below the bust of the Master, and further enhanced at the base by the effigy of a lady in a flounced dress teaching her little boy to read in one of the volumes of the Hetzel edition—has no other marks of distinction. One only finds there whatever France's botanical gardens usually offer: tulip trees, rhododendrons, swans and ducks on ponds contained by the gentle curves of well-tended lawns, like the single drop of water in the heart of a cabbage; Japanese bridges, labyrinths, and the long, continuous teardrop of water dripping down from a rock overhung by cypress and yew trees, a sight so out of character when compared with those of traditional French gardens that it looks like a forgotten relic from centuries past, a contemporary of Horace's *Bandusie* fountain. On the south side, the smoke from the Orléans train station coated the leaves of the magnolia trees with coal dust; to the east, the Botanical Gardens border lifeless, joyless neighborhoods harboring a cemetery, which crosses the silent and sorrowful rue d'Allonville. But it seems to me that on the west side, where a former wall has been replaced halfway by an iron fence that allows the odor of exotic plants to waft into the rue Stanislas Baudry on summer evenings, nothing has changed; the view on the horizon as seen four times a day from our recreational courtyard, that decorative line of dense, top-heavy greenery on the other side of the street is still the same as it was sixty years ago. That vegetal *skyline*, more suggestive for me than any profile of a city outlined against the sky, remained for many years my repertory of lines and colors, a simplistic alphabet of vegetation but inexhaustible in its combinations, and the source of emblematic illuminations for dozens of books I love. Here is where I found all those giant cypresses of the Everglades in *North against South*, the deodars of the Himalaya sheltering the hibernating house in *The House of Steam*, and, later on, the *liquidambars de la fontaine* where Atala appears to Chactas. Even today, wherever my travels take me, if I have an hour to spare in an unfamiliar city, I find myself gravitating toward those placid chlorophyllian enclaves, though so unfortunately surrounded these days by throngs of motor vehicles, towered over by high-rise buildings, and crowded in by multi-storied concrete *residences* peeking through the foliage of cedar and catalpa trees. For me, every one of those vegetal Noah's arks is a humble treasure chest, assailed from all sides, mistreated, tightly squeezed in by the tidal wave of industrial urbanization, but whose vegetal time-release mechanism will one day explode and reseed the abandoned cities. Almost everything pleases me in the similarity of their composition and layout: the lazily meandering alleys patterned after English gardens, the scalloped borders like those of croquet lawns, the wire mesh cages resembling silos stuffed with reddish autumn leaves, the beautiful Latin names of Linnaeus's flora on oval metal tags, the postman Cheval's wheelbarrow standing alone, left to its dreams, the gardeners in blue overalls working without haste at their vegetal palace, the area reserved for seedlings, sheds, and greenhouses, the stacks of empty flowerpots looking like trunks of palm trees, the little house of the curator with its crown of well-trained, flourishing greenery, its open windows breathing as easily right in the midst of a city as they would in the middle of a forest. It is Nantes who kindled my interest in these small, little-appreciated oases, and taught me how to make good use of them; a predilection which has taken me more than once back to the gardens of Avranches gently sloping down toward the sea, magnetized by the immense panorama of the shore where the minuscule cone of Mont Saint-Michel seems ready to float out to the sea at eventide, as well as to the very beautiful floral stairway of the gardens in Coutances, secluded and enclosed on all sides by palisades like a mystical garden: *hortus conclusus.*

Because of an old grudge held for a long time, I never returned to the museum, although it is located as close to the lycée as the Botanical Gardens. For me, it had been the opposite of the Gardens, the true negative pole: forced to go there like a dog on a leash because of the school's "cultural" enrichment program, I execrated paintings for a quarter of a century. Strange monument whose exhibition rooms are devoid of windows, a sort of pedestal amputated of its antique Roman chariot which brings to mind, whenever I pass in front of it, not the crypt with its stash of paintings, but rather, and I don't know why, the raised funeral temple in the kitschy film *Anthony and Cleopatra* where, after Actium, the vanquished lovers have taken refuge amidst supplies hoisted up to them by a system of pulleys positioned on several levels. It is strange how art affects us, especially official art, unpredictably and sometimes with mixed or very little results.

Toward the east, between the Erdre and the railroad, the neighborhood of forbidding and yet rather sleepy-looking administrative buildings is quite appropriately surrounded by an almost continuous string of military enclaves: the Lamoricière, Richemont, and Mellinet garrisons, and the Cambronne barracks (now the police station). That is where, in 1935, I completed my military service in the 65th company of the Signal Corps: unlike my years at the lycée, this period in my life left hardly any memories except those of Morse code exercises, carried out at the other side of the racetrack by means of a curious optical instrument, an apparatus equipped with a shutter on which many a winter rain had fallen. I don't even know if I could find again today the range at Bêle, where I went from time to time to direct mortar attack exercises. One day, when a shot had just been fired, we were surprised by the sudden appearance of a general stately galloping right into the line of fire. There were no consequences to this incident, I was not held accountable: the careless officer had been at fault. It is only behind the museum, in the cloistered, sorrowful atmosphere of the rue Gambetta, astir with the austere, morose breath of the regiment in residence but where nobody ever ventures into, that I still find traces of the city's former warlike apparatus. Quite recently, I went there looking for the mess hall where I used to lunch, but in vain; all I found was a recruiting station of the French Foreign Legion, with a sign announcing to the solitude surrounding it that their offices never close.

# EVERY OTHER WEEK,

I spent Sunday at my great-aunt's who sent her maid Angèle, smooth-skinned and rosy under her Breton *bergot*, the starched white headdress, to fetch me in the parlor after Mass. On alternate Sundays, and on Thursdays—three out of four Thursdays—my "outings" consisted of school-mandated promenades. The usual goal of these health-oriented, pre-mealtime walks was to reach an empty lot where we could play ball, in those times a more or less run-down site in some *green zone* on the edge of town. For that reason, the image of Nantes taking shape spontaneously in my mind is not a labyrinth of streets in the heart of town from which one occasionally escapes, but rather a badly tied knot of divergent radials along which the urban flow escapes and dissolves into the countryside like electricity traveling from points of distribution. This is perhaps why I am more sensitive than others to the existence of all kinds of boundaries along which the urban fabric tends to fray and unravel, areas neither within nor outside of city limits. In light of the books I have written, there are times I cannot help but think that my fondness for those borderline areas has increased since then, gained momentum, and grown to the point where, by a game of analogies, it manifests itself in unexpected domains of a more somber tonality: once we start imagining, there is just one step from boundary to frontier. In any case, every time I reread Rimbaud's poem already quoted ("The city filled with smoke and the noise of workers plying their trades followed us very far on our paths . . ."), I feel that bittersweet sensation again—almost like a dream, but which then was quite real for me—of being slightly numb, shivering, floating alongside a large, living body so close I could feel its breath, but which some evil spell kept me from embracing. Adrift on shreds of inhospitable land, slowly conquered by silence and mired in a sort of catalepsy, I could feel from afar the immense, haunting presence of the city, like that of a giant beast holed up in its lair whose respiration was the only sign of life. In almost every town where I have lived since then, whenever I went for a walk, my steps would automatically direct me toward some point of departure into the country: for example, in Caen—the Caen before it was bombed and rebuilt—after leaving my lodgings on the place Saint-Martin, I liked to take the road north to Langrune which passes through Douvres-la-Délivrande, walk past every one of the little houses lined up along the way, and continue until I had reached the vast, empty northern plains hovered over by swarms of crows, where the path cuts through the curved veins of local limestone breaking through the soil, and where a vast horizon, windy and invigorating, offered me the soothing view of the wide open country as well as the presentiment of almost having reached the seashore.

There are times when my daydreams take me even further than those recurring thoughts about the tropism of borderline areas, reminding me that the rhythm of my life and the places where I live during any given year could be interpreted by an outsider like the effects of a long hesitation between the city and the country; a hesitation never completely resolved, resulting in the compromise of my spending less than two thirds of the year in the city. It even happens that I count on my dreams at night to find a solution to this problem, something which has actually happened—in the casual, fanciful manner à la Alphonse Allais, who proposed that cities be transported into the country. I recently had one dream after another, one about Caen and one about Quimper, two cities where I lived for a considerable length of time. In the first dream, Caen was reduced to a *cours*, a wide avenue planted with linden trees, its side streets animated by a sort of flea market; at the far end, a humpbacked stone bridge overgrown by tall weeds led directly to a cow pasture—perhaps just images inspired by memories of those vast public gardens along the river Orme, the *Prairie Caennaise*. In the other dream, after leaving the center of Quimper and walking down the rue Kéréon to a path bordered by stones standing upright, directly patterned after the alignments at Carnac, it only took me a few minutes to reach a *plou*, a small Breton hamlet where a religious holiday celebration was in full swing, complete with May pole, games of hit-the-donkey, bicycle gymkhana, and a sack race.

As a matter of fact, the general image of Nantes taking shape in my mind—an ongoing, never-ending process—is a framework of radials, an image lacking a real center. The city's west side, starting from the Grillaud neighborhood inside the Chantenay district, which now extends farther north via the voluminous protuberance of Saint-Herblain—neighborhoods where not one major street originates—still reminds me of those spaces on ancient African maps left blank to designate unexplored territories: a vast, opaque zone where the blood is not circulating, though inseparable from the body of a city still very much alive in my memory. Monotonous, endless boulevards cut through this sluggish agglomeration, whose abstract names memory has already relegated to a grub-like existence: boulevard de la Solidarité—boulevard de l'Egalité—boulevard de la Liberté—boulevard de la Fraternité (this one, after a sort of hiatus, becomes the boulevard des Anglais). Enough said about these urban lands destined to remain forever fallow since there is not a single memory to fertilize them: Nantes, on the west side, does not come alive before one reaches the Chézine Valley and Procé Park.

Of all the ritual Sunday and Thursday promenades, it is the route de Vannes which has left me the fewest memories. Like the Nile in the heart of Africa, it originates in the heart of Nantes, on the place Bretagne, in the form of a winding artery swollen with unhealthy pockets such as the place Viarme. It seems to me that we used to access it only laterally, *via* the boulevard Le Lasseur, because of some school directive prohibiting passage through the rue du Marchix, then an alley of ill repute bordered by slums, a somewhat notorious hotbed for Nantes' criminal elements of the nineteen-twenties. Today, there is no part of town where I feel as lost as in the neighborhood of the place Bretagne and the rue du Marchix, completely rebuilt after being leveled by bombs, and dominated by the enormous

Tour de Bretagne standing defiantly alone, like Dracula's sharpened stake planted aggressively in the heart of that city of vampires. Confronted by the mirror-like windows of the new office buildings, I find it very difficult to conjure up the yellowish, scaly grime, the general aspect of a neighborhood that suffered years of neglect, all those houses which once formed a circle around the place Bretagne doing penance, waiting patiently to be renovated, languishing on that deserted, yawning square just two steps away from the rue du Calvaire like some county fairgrounds on a day *without* ... (memories of days *without* meat, and other wartime restrictions). There used to be—there still are—fragments of what remains of the Nantes of 1793, a scattering of houses which formed a circle around the place Viarme, where Cathelineau fell to the enemy and a small monument commemorates the execution of Charette, buildings that seem to have weathered the years. Set around the haughty, imposing town houses adorned by pediments and pilasters built by the slave traders on Feydeau Island and in the Graslin neighborhood, these one- or two-story houses with their little gardens look like the first signs of the nearby countryside. Quite a few of them have survived in the center of Cholet, a simple rural burg in 1793, but then elevated to town status in the following century like a plant going to seed without changing its foliage or odor. The vicinity of the place Viarme, close to the road to Rennes and near the rue Haute-Roche and the rue Noire, is still the best place to look for Nantes' old peripheral neighborhoods; modestly middle-class, their ties to the countryside unbroken, their ways of speech unchanged—a Nantes where, in the month of June when I went to lunch at my great-aunt's (her house only had a small flower garden), the peas and cherries on the table did not come from the market but from polite exchanges and good neighborly relations of long standing with the vegetable gardens next door.

Last year, I took a taxi to a house fondly remembered from times past; but after leaving the premises I got lost in the area around the route de Vannes, even though I passed through there so many times. I finally found myself on a long, wide, straight road roaring with traffic like a highway which I did not recognize: it was difficult to imagine that this part of town in the northwest sector, near Orvault and Saint-Herblain, had expanded several kilometers since 1925. While walking in that direction, we had never reached that straight line starting at the boulevard du *Massacre*; for us, the town ended just beyond the roundabout of Vannes, in a scattering of nondescript, isolated houses, stands of trees, and open fields. On this side of town, the ever-changing image of the god Terminus— a well-disposed, hostile, smiling, or sad deity I imagined presiding at every one of Nantes' exits—was amiss; like sugar in water, the city dissolved slowly into the country, it was impossible to tell just where it ended. Confronted by that road without any points of reference we would stop and, not knowing what to do, sit down aimlessly on its weed-covered shoulders while waiting for the signal to go back. If all the other city limits had been as unremarkable as this one, there would be nothing left today of an image which still looms large in my mind, the layout of Nantes like a stellar configuration.

Only one perfectly straight thruway, four kilometers long, crosses Nantes from one end to the other without linking up with any other road, reminiscent of some local Baron Haussmann who might have fallen victim to an inopportune change of plans at the very beginning of his urban renewal projects: this is the long street named successively rue de Strasbourg, rue Paul-Bellamy, and boulevard Robert Schumann, bordered by trees starting at the river Erdre. It leads north toward the bridge over the river Cens; going south, it stopped abruptly, without connecting to any bridge, at the northern arm of the Loire where the river divided at the tip of Feydeau Island. Because my regular Sunday outings and so many other promenades had taken me there so often, this road, enhanced by memory with a few mental drawings of adjacent, familiar sights, remains for me the true axis of the city and even more: a sort of initiatory path which became a point of departures, where many perspectives would eventually reveal their secrets, a path where I could feel the city getting closer and closer, slowly taking shape in my mind.

A few weeks ago, after an interval of fifty-five years, I again walked up to the bridge over the Cens: it was a sobering experience. Once I had passed the former riverbed of the Erdre at Port-Communeau, the houses on the street, lined up behind a double row of plane trees pruned into stumps, already looked like those of certain dilapidated, carelessly inhabited suburbs. On the left, going uphill, the familiar *gazomêtres* (cylindrical gas-pressure reducing stations once used as reservoirs) had disappeared. It is only when the wanderer approaches the orchards of the roundabout in the rue de Rennes that the street recaptures some of its charms, since in the summertime foliage and tree branches would sometimes reach over the walls to extend their greetings and—better than Mme de Noailles' roses—"make the passerby feel the benevolent season"; not all of them have fallen victim to the assault of tract home developments. But it was not just the avenue which attracted my attention while on that pilgrimage. It was the light, fresh breeze at almost every intersection, a wink from the past sent my way by every one of the lateral perspectives suddenly reopening, as if the people living there had left their peacefully slumbering houses to congregate at every intersection in order to salute my passage. All of them! They were all here! Forgotten for a very long time and resurrected intact; every open vista coming into view along that familiar walk had left its imprint on memory, like a series of snapshots taken absentmindedly on a roll of film developed at a much later date. There was the rue de Bouillé with its corner house built of bricks and white stone sitting high above the bank of the river Erdre, whose waters were more crowded with pleasure boats than a Chinese backwater. The rue Noire, with nary a dwelling in sight, haughty and sorrowful, corseted by its moss-covered supporting walls topped by tufts of bamboo, mimosa, and the branches of magnolia trees. The hollow crease of the boulevard Lasseur. The Rennes roundabout, long ago modestly enhanced by a pharmacy adorned with green and pink globes. The garden gate of wire mesh, painted olive green, which still faces the entrance to the rue Haute-Roche from the rue Bel-Air.

All these streets, absentmindedly inhabited, especially those leading east, seem to have sunk into that vegetal sleep characteristic of garden paths which slowly disappear from view. But, if one continues in that northerly direction, soon a *lighthouse* comes into view, also long forgotten though still familiar: it is the exotic silhouette of a pink-tiled roof sitting on top of a small tower—conical or hexagonal—set in a park against a backdrop of trees, which marks the spot where the street's long, steep slope turns slightly to the right and then descends to the bridge over the Cens.

Because I automatically associated it with the freedom of my Sundays exempted from the mandatory school promenades, I've always had a soft spot for the rue Bel-Air, without doubt one of the ugliest streets in Nantes. Whenever I found myself walking there on Sundays, the mechanical noises generated by young ladies pounding away on the piano poured out of the windows into the street, enveloping the passerby like some vibrating cocoon born of the sweltering summer heat, which rose like a shimmering mirage from the sun-baked road. But on the place Saint-Similien, where the rue Bel-Air turns right in front of a limestone church noticeable only because of its shameful insignificance, one of the city's strangest panoramas opens up on the south side: from the square's asphalted, steeply inclined roundabout one has a dominating view of the old Nantes, gathered into a very dark sea of slate roofs; still farther away, but looming large on the horizon, sits the *old black monster*, Claudel's *evangelical beast*, Nantes' cathedral without towers or spires, stuck like a beached whale inside the mass of houses crowding around it. As I remember, the place Saint-Similien was always closed off on the south side by a tall wooden fence which must have been put up to protect a building site; on that fence, prominently displayed, is a poster advertising *Vidocq*, the film series favored by cinema fans at that time. It showed the convict-turned-policeman dressed like the hero in *Les Mystères de Paris*, next to his clever sidekick, Coco Lacour. Why is there something so distinctly dramatic in the sudden vision of that square leaning over a sea of rooftops, especially in the winter when layers of heavy clouds are settling over Nantes? It is like a special theatrical effect immediately intercepted by the houses which realign themselves; there are no more surprises in store along the bourgeois and calm rue Jean-Jaurès which takes off to the right, crosses the rue du Marchix without being sullied by it, and then continues on to the Palace of Justice and to the prison, a hotbed of Nantes' folklore located in a fine neighborhood with good air quality, a site pleasant enough to lighten up Mr. Badinter's *Weltanschaung*.

In years past, as soon as the tramway started on its long downhill ride to the end station at the Pont-du-Cens, its tracks left the road and moved up on the unpaved sidewalk, knowing that there would be no more pedestrians to watch out for. It then rolled straight ahead, next to the wall of a park topped by a tall jumble of branches, all the way down to the bottom of the little valley. The two small open-air cafés indicating the end of the line are still there, set obliquely inside the bend, where the tracks begin to climb again. The small river Cens was once one of the city's limits, no houses ever reached that far; only a few walled-in farms surrounded by orchards clung to the southern slope of the valley.

On the other side of the bridge, at the junction marked by the church of Notre Dame de Lourdes, the road—now rue du Chanoine Poupard—climbed between pastures closed in by hedges; here stood a covered well where, on a winter afternoon, we once found a grass snake hibernating. Whenever our promenades took us along the river Cens, we usually stopped at the limestone quarries, located at the right side of the road and still operational at that time; I have tried to find them again, but could not find the slightest trace in that valley where apartment blocks and other residential buildings, set among the occasional stand of trees, now extend as far as the river. It was here, on the fallow land surrounding the quarries, where small miners' wagons stood abandoned, rusting away, that my imagination had transplanted those unidentifiable, vague sites chosen by Edgar Poe—so unlike any Parisian locale—for the setting of *Marie Roget's* murder. I could see the menacing shadows of evening settling on the scene which, according to Poe, prompted wanderers on the other side of the Seine to abandon the groves and *outposts of greenery* to the criminal activities of prowlers living at the city limits. And so this peaceful bedroom community I chanced upon in 1983 still bears the mark of an evil sign, just because at age twelve I had superimposed on the riverbanks of the Cens Poe's fanciful image of a site in the Paris neighborhood of *Roule*. It is perhaps for the same reason that my memories of the area around the bridge over the Cens are always darkened by the ashen pall of winter's dusk; it marked just the tip, the outermost edge of that sparkling coat of lights the city draped over the hill's shoulder at nighttime which stretched as far as the corner of the valley. Along the little river fished for trout, the open-air cafés have closed their shutters; everything in the lifeless pastures and empty fields has turned gray, lulled into winter's sleep. We would come back by scaling the park's shadowy, steep slope which smelled so strongly of the forest, a slope that separated the vineyards' arbors and fried-food stands from the city. To our left, white vapors were already rising to shroud the orchards and gardens which cling to the hills above the Cens.

Whereas the parc de la Gaudinière, a park where we could only guess what was hidden behind its high walls, injected a little mystery into our promenades to the Cens, the walk on the route de Paris offered only utter boredom and banality; everything along the way seemed *second-rate* to us. Shortly after Saint-Clément, we passed the façade of the Livet Professional School, a trade school in vogue at that time but universally despised by lycée students, and then turned left at the Basilica of *Les Enfants Nantais*, Donatien and Rogatien, a church disfigured by its two stumpy towers like a beast amputated of its horns.

*Minor natu Donatianus fide fit prior.*

The road continued, a desperately long, straight line bordered by houses typical of working-class suburbs, garages, small factories, and truck farmers' fields studded by bell-shaped glass lids to protect melons from damage and frost. Finally, on the other side of a railroad crossing that blocked the road, we would see rising on the left what looked like concrete aviation hangars flanked by tangles of rusty rails and cement structures supporting an overhead track-mounted crane: this was the locomotive factory of Batignolles, a desolate site where we would be dropped off for two hours, left to our own devices in an enclosed yard next to the railroad tracks which bristled with fermented hay. This entire part of the city's east side, beyond Saint-Clément, will always be for me a reminder of the dreary, polluting, rusty devastation brought about by the first industrial wave, Nantes' equivalent of Paris's grim suburbs stretching through Pantin, Aubervilliers, and Courneuve. Moreover, Nantes' bourgeois neighborhoods, where the *nouveaux riches* of the XVIII century settled after having prospered from the

commerce of sugar and ebony, failed to act while encroaching working-class suburbs kept on blocking,
and finally cut off all access to the country; not a single escape route was kept open, nothing but a narrow
stretch of land up north along the winding river Erdrc,
almost inaccessible except by boat. One still enters
the city on all sides from dreary, ugly suburbs, just
like the ones described by travelers who arrived in pre-
revolutionary Paris aboard the old mail coaches, and
were obliged to pass through the badly kept and ill-
famed entry gates of the faubourg de Flandre or the
faubourg Saint-Marcel.

Every promenade in the direction of Nantes' south
side is doubly rewarding, more than just a walk toward
the sun. There is not the slightest resemblance between the cold farmlands, the somber greenery, the
slate roofs, the lifeless villages, and the oppressive rural ambiance of the countryside which blocks the city
in on the north side like a wall, and Nantes' south side,
the *pays Nantais*—the hillsides covered by vineyards,
which could be called collectively La Haie-Fouassière,
the beautiful Rabelaisian name given to one of its villages—the sunny levees south of the Loire—the riverbanks—the open-air cafés serving white butter sauces
and frog legs—the beautiful shady spots along the

river Sèvre, the Tuscan elegance of Clisson. Nantes, growing beyond the four arms of the Loire, gained its southern foothold rather late and not without difficulty; nevertheless, it seems as though the city found its native soil there, the territory of its origins, the land perfectly attuned to its own harmony. Nantes, where people drink wine, not cider, is part of the Vendée region as well as of Brittany, though still solidly attached to and held in place by the last slopes of one of Brittany's geological formations. Venturing forth timidly toward the southern, Mediterranean-like shores of the Loire's left bank, the city seems to look longingly at the riverbanks of the villages Saint-Sébastien and Trentemoult, as if they were the frontiers of a land of milk and honey, a land from which she draws her lifeblood, which attracts and charms her, but where an unpredictable stream has denied her full access. In the nineteen-twenties, there was only one bridge, the Pirmil, which connected greater Nantes to the southern countryside: a demonstrably precarious link, since it simply collapsed one fine morning. Our walks in that direction were usually quite short: we would reach Gloriette Island from across the Lefèvre-Utile factory, a building shaped like a half-moon, topped by a brick and white stone tower sporting the initials L.U., once Nantes' major industrial icon, like the luminous Mercedes sign high above postwar Stuttgart; then, taking a short cut through the rue Fouré, we walked to the quai Magellan and the present Audibert Bridge. The large Beaulieu Island had never been more than a convenient dump site for Nantes, a Z.I.P. (Protected Industrial Zone) used as a switchyard or side yard, and to store polluting merchandise. With nothing but empty, weed-covered lots at its farthest point upstream, its downstream area crowded by forests of tall steel masts inside the huge dry docks of naval construction yards, the island was essentially a two-lane road between the bridges, a distance best traveled quickly, without stopping, as one would in a flood zone. Around 1920, it was still somewhat prestigious to live on Feydeau Island, and not dishonorable to reside on Gloriette Island, along the quai Magellan, or on the chaussée de la Madeleine—but living on Beaulieu Island could only mean that one had to toil at certain menial jobs: switchman at the Etat railroad station, or welder on a Dubigeon construction site.

At Pont-Rousseau, we usually stopped in one of the pastures bordered by pruned ash trees, alongside the Sèvre; the little river, still untouched by construction projects, flowed between trembling hedges as far as

59

the Loire. But the most charming sight of the south bank was Saint-Sébastien, a sunny village where trellises and glycinia trees shaded the narrow streets covered by old and cracked flat stepping stones. Walking along the boulevard next to the Loire, since then so oddly rebaptized boulevard des Pas Enchantés, I could already spot the *boires*, the little islands slumbering along the southern bank where willow trees grow amidst thickets, a familiar sight that reminded me of Saint-Florent. Fifteen kilometers farther down the road, the pruned trunk of the giant spruce which for such a long time had dominated the hills of La Varenne was almost visible; it had been the true signpost marking the frontier of the Anjou region above the Loire. To the east, in the direction of my parents' lands, the Loire shimmered in the light of a sky I remembered from summer vacations, casting a warm glow on this promenade which already took me halfway home; the brisk, more pressing tone of voice heard on the streets of Nantes toned down suddenly, became infused with the local dialect. Saint-Sébastien refused to be a suburb: it was an outpost of the Vendée countryside implanted on the banks of the Loire, sheltered by the river, and not in the least contaminated by the air of the city. From here, one could listen absentmindedly to the city's noises coming across the channel, like Charette's "*black sheep*," the soldiers who had witnessed the assault on Nantes in 1793 from this very spot, almost like curious onlookers. Farther west, the same frontier separates Trentemoult—a tiny hamlet with its tile-covered cottages, the *bourrines*, its narrow fishermen's alleys whitewashed with lime and hung with nets drying on trellises, very much like those found in L'Epoids, des Sables, or Croix-de-Vie—from the dark hill of Sainte-Anne, where large container ships and banana boats slowly glide by, or drop anchor.

More attractive, and richer in images than those boring, aimless promenades taking us into areas where roads turned into highways leading to Vannes, Rennes, Paris, or La Rochelle, were the walks leading to dead ends at the city limits, across green zones where only half of the area was cultivated or maintained by the city, like some of London's *commons*. There were still a few left then in Nantes; one side connected to the city and well maintained, the opposite side neglected and already part of the adjoining countryside. Several years later I came across a much larger version in London's *Hampstead Heath*, very close to the street where I lived; it started out as a botanical garden on the city side, but then, after being gradually invaded by a sort of horticultural *fog* which obliterated its stands of trees, shrubbery, bowers, and paths, it grew completely wild and finally turned into a true Scottish heath at its northernmost edge. This transitional process was not without its utilitarian side; at a time when English hotels still refused to rent rooms to unmarried couples, the terrain's medium-height bushes and arbors provided an alternative refuge for *off-the-record* trysts in the suburbs.

What I enjoyed finding again in Hampstead Heath, during those twilight hours when lovers furtively meet on summer evenings, was the image of a gradual transformation affording an escape from trodden paths, a vision of the fascinating disorder taking hold of a landscape when it is no longer clearly defined. When it conceals, and at the same time authorizes, deviations from the generally accepted norms in the comportment of those wandering on its grounds. The expression *terrains vagues*, to which I paid tribute elsewhere, stands here for both a desire and a favorite image: the confusion which clouds here and there the city's boundaries transforms them into dreamlands, as well as places where one is free to roam. For me, Nantes' *terrains vagues* were le Petit Port, la Colinière, and Procé Park—and even though the time we could spend there was limited, at least nothing could constrain our imagination.

★

Looking back, few of our promenades' itineraries seemed as magically empowered to evoke people and places as the one Breton himself followed so often in 1915-16 and later wrote about, between the girls' lycée in the rue du Bocage, at that time a military hospital, and Procé Park.

Walking the streets of Nantes, I am in the thrall of Rimbaud: what he has seen, far away from here, interferes with and even substitutes itself for what I see; to put it in his words, I have never lived that kind of "double experience" since then. The rather long path which takes me every afternoon, alone and on foot, from the hospital in the rue du Bocage to the beautiful Procé Park opens all kinds of perspectives evoked in *The Illuminations*: here, the general's house in *Enfance*, there "that hunchbacked wooden bridge," still farther on, some very strange happenings described by Rimbaud: all of which vanishes into a certain bend of the little river surrounding the park, which becomes the "river of cassis." I cannot give a more coherent idea of those things.

What strikes me in this passage, where certain parts touch a chord deep inside of me, is not just the superb accolade to Rimbaud's power of descriptive

6 3

imagery, but also Breton's tribute to the particular aptitude of a city to endlessly and effortlessly provide landmarks, paths, and models for an imagination seeking poetic inspiration—an experience relived, and verified in later years, under different circumstances —which explain or justify the most unusual visions almost matter-of-factly, without being prompted. Though less often than Breton, I have also followed the easily identifiable itinerary at regular intervals, starting at the yellow brick lycée in the rue du Bocage (no doubt the twin of the lycée Clemenceau although the latter was built of carved stone, a distinction the scholarly *machismo* at the beginning of the century reserved for the boys' lycée), passing through the rue Mondésir and the rue de la Bastille before reaching the park via the rue Dervallières. Even today I love to walk along the rue Dervallières, which seems to be on the brink of being abandoned, untouched by any kind of renovation, or at least noticeably older than anything around. On the north side, moss-eaten walls leaning into the street are straining to contain the luxuriant growth of ancient parks grown wild, walls weighed down by overhanging branches—only the minuscule residential complex *Toutes Joies*, located a little farther east, a neat and placid-looking enclave resembling a Dutch settlement with its gabled villas and sand-covered paths behind the iron gate, brightens this street like a patch of blue sky after a storm, a street which seems to be of the same age and appearance as the one adjacent to the rue Plumet's gardens in *Les Misérables*.

Procé Park as it looks today—surrounded by an enclosure, its lawns trimmed, its alleys spread with sand, raked and well kept—fulfills the expectations of a modern *green zone*; but it is hardly the park Breton knew, nor the one I discovered in the early nineteen-twenties, no doubt still unchanged at that time. A plaque affixed to the gate, on the side of the bridge over the river Chézine, indicates that the land had been donated to the city in 1912 (when Paul Bellamy was its mayor) by a certain Mr. Caille. There were no houses in view at that time, except on the south side; today, towers flanked by balconies rise behind the beautiful stands of conifers at the east-side entry, quite visible through the branches. The park was once wide open, and half-wild; years later, visitors and nearby residents no doubt welcomed seeing it gated and patrolled. But if, on a fine clear afternoon, one walks all the way to the brick bridge with the narrow arches, built like an aqueduct, which spans the river Chézine and marks the end of the park on the west side, there is a surprise in store: looking through one of the

arches, one is rewarded with a *view* framed with un-
canny precision, just as disorienting and startling as
the miniature photographs inserted into the pens
schoolboys used in my childhood. On the far side of
the bridge, bathed in the yellow light of the setting
sun, a little valley spreads out its fields and pastures;
its rather steep slopes are closed off on each side by a
row of solid thickets—just as solitary, as uninhabited,
and as silent as a valley in the Ardennes Forest. And,
for just a few more months—perhaps only until the
time when, as announced by a municipal poster, the
development of the "val de Chézine" will be completed—Breton's vision remains intact: ". . . the river
of cassis still flows here, observed by no one, through
valleys unknown."

Unlike the important thoroughfares which almost in-
variably encourage urban settlements to spring up
next to them and grow far into the countryside, the
Chézine, flowing down from the northwest toward
the heart of Nantes, pulls alongside its riverbed a
green trail of dewy meadows, interrupted only here
and there by the mosaic of small settlements; it continues, almost undiminished, as far as the rue de Gigant where its waters disappear underground at a dis-

tance of less than one kilometer from the Loire. Still almost completely intact next to the Dervallières stadium, developed and maintained along Procé Park, this vein of greenery penetrates into the city in a sprinkle of lawns, stands of trees, gardens, and playing fields. This stubborn refusal of such a little river to capitulate until the very last moment when a mass of stone and cement finally closes in on it, the bouquets of fresh foliage it still waves from afar in the wind above the roofs before coming to a stagnating stop at the sewer drain, are part of the charm of the low-lying gardens beyond the avenue Camus which this small body of water keeps alive and beautiful. Whatever the Grange Batelièrc in the heart of Paris signified a long time ago, or, in the last century, the Bièvre when it powered the mill of the *Moulin des Prés* and accompanied its ribbon of pastureland all the way down to the Seine, here in the midst of Nantes it is still the Chézine which halfheartedly succeeds in bringing back memories of times when the street and the stable, the workshop and the pasture still went hand in hand, when one could enjoy Sunday lunch on a grassy knoll without leaving the city. I still feel nostalgic for these times when I visit Amsterdam and see fishermen right in the heart of the city who tend their lines in a canal at eventide, close to their calm, peaceful brick houses, and then stretch out full length on the new grass along the water's edge.

Few itineraries are as deserted as the one leading, via the boulevard Michelet, to le Petit Port. On the right, where the boulevard Amiral-Courbet bends, just past the bell tower of Saint-Felix—one of the ugliest among the city's many ugly churches—a straight, almost un-inhabited avenue comes into view, mournful and sad under its sparse foliage of plane trees pruned into mu-tilated stumps, something only a population unfamil-iar with vegetation in the south of France could con-sider a pleasant sight. The *Loquidy*, a lycée and its park difficult to visualize behind the prison-like wall, extends along the left side; on the right—and now the impression of barren desolation which has settled on this avenue like a layer of chalk-like dust becomes even stronger as we approach it—lies the Morrho-nière bus terminal, formerly the tramway terminal, stretched out underneath a bellow-shaped roofline and a watchtower of hexagonal tiles perched high atop its offices. Every type of dereliction found along a city's beltways, where from very early morning on the incoming traffic seems to include the day's share of pain and sorrow, marks this avenue so profoundly that while returning to the post from our field exercises, the band of the 65th regiment would spontaneously start playing a cheerful marching song to chase away the putrid exhalations of such a comatose perspective.

Nevertheless, that gloomy, depressing road has a surprise or two in store. Right behind the Morrhonière, after crossing the valley of the Cens which deepens just before the Cens flows into the Erdre—a valley still undeveloped then, but now converted into public parklands—one enters a green zone which has not yet found its calling; neither a public park, although the arrangements of trees growing between the clearings still evoke some rudimentary ornamental layout, nor a private domain, since there are no fences. Rather like the grounds of a castle no longer maintained after the main building has been destroyed, a domain returning to its original state of simple pastureland, where all the signs of the former perspectives, corners, and angles have almost disappeared. Nevertheless, a hint of past grandeur still lingers on, an elegance inherent in the placement of its groves, and in the refusal of its lawns and arbors to serve any kind of purpose.

Beyond the little valley of Cens, where the road to the racetrack starts to climb again beneath the trees, it passes a building on the right set on top of a low hill, surrounded by a wall; a place where, long ago, I spent hours lost in dreams. Only half-light filters through the branches of a dense ring of trees surrounding the wall which isolates the building; darkness never ceases to hover over that remote domain like the shadow of a rain cloud. According to city maps, this is the location of the *Ancien Observatoire*, heavily protected in times past by thickets of nettles on the north side; inside the wall, there is a windowed structure topped by a crudely fashioned, square protuberance from which two or three archaic measuring instruments—one of them looks like an anemometer—point at the sky. I don't know if it was because of the decrepit, outmoded instruments—so much like those handled by the phantom crew of the *Manuscript found in a bottle*—or rather because it looked very much like an inaccessible cloister, apparently uninhabited although still marginally kept up, but the moment I chanced upon this sadly neglected site, I was fascinated: the former observatory became my point of reference and central image needed to visualize the house of Usher, Dracula's castle, all the haunted houses depicted in novels or paintings like "the house of the hanged man." The ivy, the water-saturated moss of the wall surrounding the enclave, *the evil eye*

ambiance, the sinister vigil of the building's windows
piercing the darkness about to settle around the hill,
are not only images and impressions I projected mentally, year after year, on hundreds of pages in the books
I read; they have come back to me, intact and unchanged, in situations most unlikely and at moments
when I least expected them. For example, when, in
*Port Blanc*, on an already deserted September night, I
saw a summer cottage with closed shutters project a
powerful, unexplainable ray of green light from its
gable, almost like the beam of a lighthouse; or, while
walking late at night in *Vasterival* on the path leading
to the hotel des Terrasses, a place I fondly remember,
when I noticed a very bright light on the façade of the
first villa I passed which illuminated a sign with an
unexpected, alchemic name: *L'Athanor*. How is one to
explain the power of these unexpected objects and encounters, which instantly assume their places where
memory and imagination intersect, and take control
of the mechanism that projects a material image on
some vague memory or text only indirectly evoked?
I am inclined to believe that almost all of them are
powerful, exemplary figures endowed with multiple
meanings, and therefore creators of force fields which
magnetize everything that comes close to it: emblems
of a secret science, evil or perverse, as in the case of the

7 1

*Ancien Observatoire*, which joins forces with the passive malice typical of the closed-in site, the *haunted mansion*.

I returned to le Petit Port several weeks ago. Nothing had changed over the last sixty years at the *Ancien Observatoire*, neither the spongy walls, nor the hostile silence of the dwelling, nor the thicket of nettles; the rusted instruments protruding from the observation station still interrogate the poisonous shadows cast by the trees. And the ancient, icy wave, which seems to maintain that domain in a state of hibernation—a building doomed to fall into ruin ages ago, together with its rampart of branches still intact—again swept over me.

The last stop on our promenades in that direction was the racetrack of le Petit Port, which changed considerably over the last fifty years. It seems as though the grandstands have been rebuilt and enlarged, the track is now twice as big, fenced in, and partially covered with cinders; the entire complex looks neat and carefully maintained, an impression hardly ever conveyed by the old racetrack. It looked so neglected, so sadly abandoned during the off-season, an unused facility standing there quite empty, trembling under the cold winter rains. Behind the soaking wet grandstands, a ghost-like crowd of elegant patrons, chilled to the bone after the jockeys' ceremonial *weigh-in*, beat a hasty retreat among the leafless arbors and behind the opaque screen of magnolia trees. On the lawn, easily reached by stepping over the finish line, a single wire held in place by posts, stood a solitary shed covered by corrugated metal roofing whose partitions were removed during the off-season; it must have served as shelter for the betting windows in the summertime. Then as now, the route des Tribunes runs behind the racetrack; but on the north and west sides, there was only the wide open country. Heather and patches of broom had invaded part of the lawn's north side; beyond the racetrack, bordered by a hedge and rather poorly staked out, only a meager, stunted vegetation covered the soil: small grasslands with hedges infested by a profusion of severely pruned dwarf oaks. It seems to me that rain fell constantly, mercilessly on that terrain, whose fields of broom and heather under a vague, empty horizon continue to provide the setting in my memories for a sentence in *René* ever since I first read it: "During the daytime, I wandered far into the great heaths, which would become forests." Even today, it seems to me as if that racetrack slowly devoured by the moors still evokes for me an image more powerful than that of sheer solitude, suggested by the idea of a gradual transformation in *René*'s sentence: depopulated, isolated areas are the forerunners of a land of absolute solitude.

Nearly half a century later, my memories of the racetrack at le Petit Port gave rise to the story of King Cophetua. In order to understand its meaning, one needs to remember that in the nineteen-twenties—a time when *mail coaches* outfitted and harnessed according to British fashions still drove up the Champs-Elysées on the day of *la Journée des Drags*—the idea of ritual elegance inside the *enchanted circle*, so powerful in the minds of youths, especially when it is beyond their reach, was still very strongly linked to the English ceremonial of the jockey's weigh-in, to images of gentlemen wearing pearl-gray top hats, the *Royal Box* at Ascot, and the Day of the Grand Prix in Paris, high society's signal of departure for summer vacation. An echo of the splendor attributed to these inaccessible worldly pleasures still clung to the arbors of le Petit Port's racetrack, soaked by the winter rains. And, just as the arbors of Compiègne and the poplar trees along the Marne are still able to conjure up a gentleman's velour jacket painted by Manet or a hoop skirt by Winterhalter, because props and settings have become inseparable from their works, it is the link established during childhood that made me think of it as

**7 4**

the image of la Belle Epoque and its lost elegance, like the splendor of a Château d'Enghien or Chantilly battered simultaneously by autumn rainstorms and the ravages of war. While describing the villa La Fougeraie, I resurrected all the imagery of its drenched isolation from the pitiful shadows cast by le Petit Port. Books have their roots, like plants; and just like those of plants, they are often without grace, and without color.

It is odd that these various dead ends in Nantes' suburbs, places where we would pause on our promenades, have remained in my memory without being connected topographically: tightly sealed enclaves, cut off from each other, inserted along the edges of the city, inaccessible except from its center, with no direct links between them. Many years later, I was confused by the discovery that one could walk from le Petit Port to the bridge at Cens in fifteen minutes without having to hurry. These domineering imaginative models, stereotypes which impose themselves early on in childhood, are known to influence our reading habits and color our dreams; they must remain separate and unconnected, or they lose their power of attraction. As Proust has so excellently pointed out a propos of the "ways" at Combray—

Swann's way, the Guermantes' way, the Méseglise way—every time, each of them is reached directly, separately, after starting out from the little town, without a single allusion being made to the possibility of peripheral connections between them. This taboo cares little for geography's contingencies, but remains watchful that their primary, affective function be preserved, something which in turns depends on their location—they must be perceived as blind alleys, and remain completely isolated.

Aside from certain class periods of passionate interest to me, the only really pleasant memories of my years at the lycée are linked to a few summer Sundays (every other Sunday) when, unable to *go out,* I had to join my comrades on our collective promenades. On those Sundays, when the weather was really beautiful, the boarding students not rescued for the day by family or friends would leave in the morning right after Mass and then walk, in pairs, to La Colinière; a small rural estate in the suburbs which I believe someone had donated to the lycée. A carriage loaded with food, drink, dishes, and cutlery preceded us, together with the cooks. After reaching the Doulon railroad crossing we would turn right on the little road to Sainte-Luce, and, after a lengthy, winding walk between the greenhouses of truck farms, little gardens, weekend cabins, and small open-air cafés open only on Sundays, located farther and farther apart, our double file would break up into small groups; heads went uncovered, jackets were unbuttoned. By the time we reached the high grass of the country meadows abloom with flowers, the odors of the sugary, overheated sap rising from the earth in June went to our heads like a foretaste of summer vacation. About a kilometer down the road, just a few orchards away, we would glimpse the minuscule village of La Colinière, almost hidden within tall stands of summer's feathery growth, to the right of the narrow white road (asphalt was then still unknown); and, on the left, the lycée's country estate, set behind a rusty gate that opened into a walled-in, grass-covered courtyard heavily shaded by trees planted at random. To the left of the building, behind a low wall, stood the tenant farmer's house; in front of it, a very large, lawn-covered courtyard—a sort of mall which looked like an extension of the front yard, planted with ash, maples, and acacias, closed off on both sides by thickets—ran all the way to the railroad tracks (it must have been the Segré line) hidden behind a hedge. It was not difficult to cut across it so that we could—I no longer know for what purpose—put bronze coins on the tracks to have them flattened by passing freight trains. The grounds where we were allowed to play—the central courtyard and two adjoining yards enclosed by walls, hedgerows growing wild, bramble and briar patches—already seemed part of the farmlands; supervision, away from its usual surroundings, became relaxed; the most daring among us stepped over the low wall to strike up conversations with the tenant farmer and his household, and sometimes ventured as far as the sleepy village which seemed to make its restful, lazy morning hours last throughout the entire day.

It was difficult to guess what could have been the purpose of the main building before its donation to an educational institution; without doubt not old enough to have been an authentic country manor, and, since its rooms were large enough to accommodate sixty people for a sit-down dinner, too unwieldy to have served as a bourgeois summer residence. Its silhouette comes back to me every time I reread the episode of *La Vivetière* in Balzac's *Les Chouans*; I can see its walls rising behind fringes of tall grass, the entrance steps of rough-hewn limestone, the small, square tower linking the two wings set at right angles, the tall, latticed windows which apparently had never known curtains. Though far from being a ruin, and even adequately kept up, it was a lifeless building. Oc-

7 8

cupied only on a few Sundays during the height of the summer season, when the penetrating odor of wormy wood invaded its high-ceilinged rooms and clumps of moss swelled up between the stones on the entrance steps, it never lost that feeling of twilight settling in during long, uninhabited winters; even the strong summer sun could not eliminate its air of humid confinement. Unfurnished, except for dining tables of rough-hewn wood resting on trestles and refectory benches, bathed in an ancient, fading light, it seemed to consist only of many spacious, cleanly swept attics, reminiscent of those summer homes set in farmlands or vineyards known as *campagnes et vignes* in centuries past; places buried all winter long underneath dead branches and rotten leaves, refurbished by stewards who would arrive with horse-drawn carts carrying furniture and draperies to prepare them for a few weeks' occupancy during the summer, without really disturbing them in their sleep.

At noon, a lunch of cold meat, salad, dairy products, and fruit was served on the rustic tables of the great hall; at four o'clock, it was pâté sandwiches with a glass of weak iced coffee. On the south side, green foliage brushed against windowpanes made of old-fashioned bottle glass; looking through windows on the north side, one could see the grass in the courtyard and observe the leisurely activities in a farmyard spruced up for Sundays. I remember village municipal offices lodged in abandoned or unclaimed country manors where I saw the same kind of whitewashed walls, exposed beams, rooms with wooden floors sprinkled *en huit* (floors cleansed by first sprinkling them with watering cans, leaving trails of figure eights on the floorboards), floors still strewn with dead flies from the previous summer, windowpanes held in their frames with cracked, chocolate-brown putty—interiors just like those at La Colinière, bathed in the same atmosphere of administrative dilapidation, of a court-ordered seizure waiting to be executed, or of a household move not yet completed. But that careless decay, that anticipatory feeling of a long summer's *farniente* made us happy: La Colinière was a stop toward the end on that yearlong slope where things were slowing down, and the long summer vacation just around the corner; almost like an intermediary station at one of the levels of decompression where deep-sea divers pause briefly before coming up for air. Still subject to regimentation, but in a relaxed, Sunday-like atmosphere, La Colinière was already an integral part of the countryside, basking in the sunlight and alive with the chirping and humming of birds and bees. That smell of springtime suddenly uncorked under our noses like a bottle of strong wine, a Rimbaudian spring exploding before our eyes like a thunderstorm, left us intoxicated: we played cards, stretched out on the tall grass like Manet's picnickers, we slit the bark of maple trees with our pocket knives to see the sticky sap run: its sweet, slightly nauseating taste was like a drug, an exotic liquor that makes one's head spin, and dream of traveling to faraway places.

We would make our way back in the amber light of early evening, leave the smell of the new hay behind us, and reach town at that bewitching hour when the sweltering streets and the depressing boulevard de Doulon are set ablaze by the sun setting behind La Fosse. I cannot remember ever having seen La Colinière in the rain; and it is only in certain Russian novels that I still find passages evoking that powerful layer of heat which starts to settle over the earth at nine o'clock in the morning, that *solid rock of air*, and, underneath, those lush green islands of saturated vegetation spreading like cool, fresh ponds, signs of the earth renewing itself with sap drawn deeply from within.

La Colinière is another place I never returned to. To-day, seen from the train, it is very difficult to distinguish the bell tower from what was once the village, now lost in the rising tide of suburban housing developments which reach almost as far as Sainte-Luce. The country house, that modest institutional *folly* which served as the setting for our summer Sundays, is now surrounded on all sides by the efforts of urbanization and has grown into a full-fledged lycée; I would never be able to find my way again in that countryside once overflowing with the pungent, vegetal dampness of June days, but now asphalted, cemented. The outdated images I still keep of it are secretly dedicated to the god Pan, and to a certain state of intoxication where the fermentation of puberty and that of the earth mingle, and become one.

# I DON'T BELIEVE THAT

Balzac was particularly interested in Nantes (which he must have visited at the time he discovered Guérande, and planned *Béatrix*). A city too restless, too adventurous for a novelist who preferred to study urban life—with the exception of Paris—as reflected in the stagnant waters of Saumur or Limoges, Alençon or Angoulême, and who never set foot in Marseille, Rouen, Lyon, or Bordeaux. But there exists a "Balzacian" Nantes—most of it now a pedestrian zone—concentrated in a small area which no doubt corresponds *grosso modo* to what was the original core of the city. Its boundaries are the Château to the east, the former riverbed of the Erdre in the west, one of the Loire's former eddies in the south, the rue de la Marne and the rue de la Barillerie up north. It is a maze of old, narrow, winding little streets around the Marché des Innocents, similar to the Parisian neighborhood famous for its riots, an inner city area that must have felt the heartbeat of its populace for a long time. The French Revolution had installed its guillotine there in 1793, and Carrier, the revolutionary leader of the *Convention*, his club *Vincent La Montagne*; it is also the neighborhood where, in 1832, the Duchess of Berry found a hiding place in the home of the Misses Guiny, right behind the fireplace screen (Balzac himself could not have done better). A brisk current of ancient, inexplicable passions, never quite put to rest, still chills the air in the small, windswept intersections around the Bouffay and the church of Saint-Croix, a reminder of Nantes' tendency to outdo all other cities when blowing off its share of national political steam, in 1793 as well as in 1968. But this historical radioactive impregnation, which might intrigue an older person wandering through the streets of an unknown city, holds no interest for a child; and, oddly enough, Nantes' very old neighborhoods never have attracted me: neither the rue de la Juiverie, nor the rues de la Bâclerie, des Echevins, du Petit-Bacchus, or de l'Ancienne-Monnaie—sites of historical excavations favored by Nantes' archeologists, which I crossed without paying attention to them or often avoided—ever made me dream.

The Nantes which speaks to me, which always spoke to me directly, begins in the eighteenth century. There are very few other cities which give the impression of having rid themselves of the vestiges of the Dark Ages as resolutely, and as brutally, as Nantes right from the beginning of the Age of Enlightenment;

there are hardly any medieval buildings like the *maisons d'Adam* in Anger, and hardly any exposed cross beams, sharply pointed gables, or examples of corbelled construction. Neither the Château nor the cathedral is still encumbered by those ancient fragments of feudal or clerical medieval structures, appendages which in Angers provide the picturesque substance of almost an entire neighborhood. Just like in the old Parisian neighborhoods—where only a compact mass of edifices solidly wedged together amidst narrow, winding streets still survives from the age of barricades—it is the city map rather than the shape and outward appearance of Nantes' buildings that can take us back to centuries past; an abstract return since there is no tangible evidence, all that remains are archeological sketches which cannot be animated by superimposing images of actual life. Trying to recreate the flow of traffic in times past by establishing a pedestrian zone with footpaths seems to be as futile as setting up a model network of ancient roads, streets, and alleys in some Musée Grévin.

The neighborhood extending farther north, between the cours Saint-André, rue de Verdun, rue de la Marne, and the cours des Cinquante Otages has always intrigued me more. Though less densely settled and more spacious, it is far less lively, in spite of its proximity to the overland bus terminal and to the multitude of municipal agencies recently implanted around City Hall. Cold, shadowy, lifeless streets abound, which pass here and there in front of old town houses with a rather proud mien, though less impressive than those built along the wide cours. They silently shed their paint in the humid twilight, looking like exiles in secondhand clothes reduced to living in a run-down neighborhood; one has the impression of visiting a part of town once swank, even aristocratic, which fell on hard times and then became half-abandoned; one of the streets so sadly walled in on both sides by encroaching decay is called the rue du Marais. Still a residential area, but suited especially for widowhood or retirement; gusts of wind chasing dead leaves over the walls of the occasional, tightly locked garden, rain pounding on the zinc roofs are noisier here than elsewhere. To settle in this neighborhood is to resign oneself to the fact that life, instead of following a rhythm of daily activities, henceforth becomes regulated by the ringing of church bells which lazily count the slowly passing hours: gentle, cautious reminders of the church canon, every stroke calling to mind the intermediary void between meditation and yawning.

*

On the other side of the Erdre's former riverbed, which has now become the wide avenue cours des Cinquante Otages, lies the Graslin neighborhood, named logically after its theater; it is the center of a perfectly autonomous, second nervous system in the heart of the city. I know of no other city in France where, from morning to nightfall, the theater casts such an imperious, long shadow over the entire neighborhood gathered around it in a dense constellation of streets that carry religiously venerated names—rue Crébillon, rue Voltaire, rue Jean-Jacques Rousseau, rue Grétry, rue Piron, rue Regnard, rue Rameau, rue Le Kain, rue Racine, rue Corneille, rue Molière, rue La Fontaine, rue Scribe, rue Boileau, rue Gresset, rue Marivaux, rue Le Sage. Built during the Age of Enlightenment for the slave traders' *cultural needs*, it is still a powerful reminder—or rather used to be, half a century ago—of how theater became the pinnacle of artistic expression in the eighteenth century because of its ability to convey human emotions, a secular cathedral consecrated to an art which continued to reign supreme throughout the nineteenth century. A cult that became the direct rival of and anathema to the Church which analyzed and condemned it in a manner most lucid, almost as richly provided for in regard to liturgy

8 7

and rituals, better attuned to fluctuations in public sensitivity, and more receptive to the need for change —a place of worship where Stendhal was without doubt its most devoted and most exemplary parishioner, at least in our literature.

How can I explain why those ancient, plush crimson seats, those gilded ornaments of the Graslin theater have always cast for me such an unmistakable glow on the entire complex of streets that surround it? Unforgettable to the point where, even today, when I walk up the entrance ramp on the rue Crébillon I still have the vague feeling of stepping on a luxurious red carpet rolled out before its stately columned temple. Another reason might be the sight of the entry to the cours Cambronne located almost opposite on the square, jealously guarded by its gates, an avenue laid out like a true residential theater, where terraced houses scrutinize each other across a mall planted with flowers like theater stalls and boxes on opposite sides of the orchestra section; two narrow streets, the rue de l'Hérronnière and the rue Gresset, run parallel on each side of the cours. Memorable also because of the area behind the theater where, around 1920, the rue Scribe provided a kind of popularized, coarse echo to Graslin's *Grand Théâtre* in the form of a little variety theater. The street has since changed its ambiance and become a pedestrian zone; but at that time it was the scene of vaguely clandestine comings and goings, bathed in an ambiguous twilight of flirtatious exchanges, a seemingly more dubious, more *louche* extension of the artists' dressing rooms in the rear of the Graslin theater. During the nineteen-twenties, its reputation of a rather opulent place of rendez-vous transformed the rue Scribe into a behind-the-scenes area of discreet procuring activities, the opposite pole or reverse side (though inseparable) of the theater's neoclassical façade and columns inscribed with the immortal names of dramatic and lyric art. This intimate connection, inherent in the site itself, between the profound exaltation I experienced while at the opera and the fascination-repulsion emanating from a realm which I suspected to harbor the crudest form of eroticism made the Graslin neighborhood the veritable hot spot of the city in the eyes of an adolescent, a zone of high tension electrified by its opposite poles which provided a striking contrast to the lethargy and quasi-comatose torpor of almost all the peripheral neighborhoods.

In order to find that atmosphere again, something no doubt more than half-invented by me, although the scenery was quite real, I only need to stop for a moment in the café Molière at the corner of the rue Corneille and the rue Racine. Is it the *mana*, the evocative force generated by this illustrious trio which attracts a little colony of secondary school and lycée students in blue jeans who congregate here on school days like clusters of seashells on a reef? I don't know why or how, but the Molière has indeed succeeded, by renewing its customers, to remain an *in* place across the years and even the generations. Stendhal's *Mémoires d'un Touriste* reminds us that he liked to breakfast here while in Nantes; today, when I look at these adolescents with big hair, I see youths of the first quarter of the century, sons of families prominent in naval construction or the canning industry, hair slicked down and well groomed, wearing spats and stiff collars, sitting in front of one of those *pyrogènes* dear to Apollinaire (an ashtray equipped with a flint stone to strike a match, subject of the poem *Le pyrogène à cheveux rouges*) while waiting for their first date with some little theater rat. From behind the big plate glass windows, diagonally across and slightly to the left of the square, the cours Cambronne with its magnolia trees comes into view, well protected behind its iron gates; still farther to the left sits that fancy candy box from the Belle Epoque, the restaurant La Cigale. Just like at the brasserie Lipp or at Bofinger's (although reduced to a smaller, provincial scale), its décor of ceramics, its arabesques reminiscent of an old métro station's entrance, and its sensuous lines as seen in posters drawn by Capiello or Mucha have remained unchanged. Directly across from the café Molière, memory carves out the porch of the former hotel de France with its lobby suffocated by potted plants and discreetly shaded reception desk from what is now a plain, bricked-in wall, and evokes the suicide of Jacques Vaché, like a surrealist ribbon in a buttonhole since then sewn shut. To the extreme left, on the other side of the rue Corneille, looms the smoke-blackened side wall of the theater with its iron guard rails and rusty zigzags of fire escapes. Even today, the air one breathes on that narrow esplanade is not quite the same as the one other people around us are breathing—all it takes is imagination, and the right frame of mind.

It was not just the monumental physical presence of the theater which animated the public spaces of the entire neighborhood; all day long, first and second tenors, bassos, romantic leads, comics, baritones, ballerinas, orchestra conductor, all those in its employ, hired for the year but whoses contracts were usually renewed, an entire clergy of lyrical artists organized into a rigid hierarchy would meet every day in groups for rehearsals, and then spend the rest of the day strolling the streets around the rue Crébillon, a parade of bluish chins and profiles who dreamed of being immortalized on medals while they themselves were often the objects of dreams. More than the cinema, it was television, and then, half a century ago, the final dismantling of the last lyrical provincial ensembles engaged for the year, which caused the disappearance of these curious types of semi-celebrities found among local theatrical groups. Familiar figures because one would run into them in the café or at the newsstand, and nevertheless sublimated because of their special status which set them apart, like the sight of a priest in a sports coat, a vision quickly adjusted by our inner eye to show him wearing his robe and carrying the sacred ointments. A handful of these individuals worshipped from afar thus participated capriciously in daily life—some of them in a manner quite down to earth, by becoming involved in elopements or embroiled in conjugal tempests—they appeared and disappeared from the scene, sometimes identified, sometimes unrecognized like those disguised angels in the Bible. Apparitions with consequences entirely disproportionate to their small number, like the action of a few drops of yeast in bread dough; at that time there was no segregation, no artificial distance which cause today's star to become not just an object of fixation, but a cause of sheer frustration for the masses. It is the Nantes of 1920, living at a pace that resembled the rhythmic breathing of its lyrical scene (although the quality of voices and performances must have been modest), which immediately puts me on the same wave length as Stendhal, the passionate pilgrim to Italy's theaters, whenever I reread him today (all the theaters he mentions were located in little towns: Rome then had a population of 100,000); I can understand the ecstasy he felt when he first saw *Matrimonio segreto* (in the theater of Ivrea, I think). Behind the image of the gap-toothed *prima donna* who caused him such delights, I see the pallid face of the tenor *Farini*, gassed during the First World War; it was rumored that he had only one lung, but the legend surrounding him fired up the public's imagination into believing that the lyrical message filtered through the damaged organ in an even purer form, proof that audiences were ready and willing to go more than halfway in their support.

Throughout my life, nothing could ever diminish the prestige of the opera which, to a large extent, had been built up right from the beginning by the *aura* cast over an entire neighborhood dear to my heart. Whereas the fine arts museum, as I already mentioned, was an unfortunate reminder of the school's proximity and despised as part of our curriculum, the opera remained unscathed, free of any kind of pedagogical blemishes, and thus came to represent very quickly—splendid in its isolation—the only refuge from "real life." A radical inaptitude to assimilate even the most elementary aspects of musical technique, confirmed over a number of years, only increased its prestige in my eyes. Concerning the matter of lyrical theater, I am incapable of a well-founded judgment since I lack the necessary critical, theoretical knowledge: I am inclined to reject everything at once, or completely fall for its charms. And if, after leaving Nantes—even though the light cast on the city by the years I spent there remains unchanged—I put a lot of stock into literature, I could never concede to it the power of causing an incomparable, extraordinay breakthrough, of being able to completely captivate an individual's sensitivity, something which opera made me feel in its truly magic moments. This *otherness*, sorely needed by an imagination hounded by the aridity of the academic regime, has never stopped to confer its magic touch on the Graslin neighborhood whenever I visit it again.

It is odd that the passage *Pommeraye*, which remains the most noteworthy landmark of the neighborhood, a sight known to draw visitors not forewarned almost instantaneously into a dream world (with André Pieyre de Mandiargues at the top of the list), does not figure more prominently in that half-inhabited, half-dreamed-of imaginary landscape taking shape during my haphazard explorations of the city. A city's power of seduction linked to its "passages" has erotic affinities which are structural, and obvious: obsession with orifices and all the secret, warm, shadowy passages leading to the visceral labyrinth, the intimate recesses of the vast urban body. It is the imagination, much more than the eye, which places these shops and businesses of an inner city commercial district in a setting lit only by the dimmed lights found in bedrooms and alcoves (it is interesting to note that all those without a natural complicity with feminine secrets have excluded themselves: one finds furriers, jewelers, hairdressers, florists, shops selling shoes and gloves, but rarely a hardware store, pharmacy, or grocery). Even while venturing into newly constructed passages like the one now joining the rue de Sèvres and the rue du Cherche Midi, I hardly ever pass through them without feeling that slightly clandestine charm found in a secretly erotic *souk*: one's pace slows down involuntarily, while the eye probes the boutiques bathed in twilight as if they were the compartments of an aquarium, where sometimes an occasional shadow moves about languidly. All those compartments, slightly darkened as if protected by blinds, are completely visible from the central walkway through the glass panes of their shop windows; they represent the ambiguous spaces of some indoor or open market venue where the silent, cautious trading of objects which takes place in the strictest of privacy seems but a pretext, a cover-up for another kind of exchange more subtly regulated, and more voluptuous. Besides, one has the feeling that a closer relationship unites the men and women in charge of these grottos than the one usually found among merchants on a street with adjoining shops: one catches them visiting each other in their boutiques, engaging in conversations with voices kept low, and carrying on as familiarly as water carriers and shower attendants gossiping in the subterranean regions of a thermal bath establishment.

For me, the passage Pommeraye lacks the secret attribute which renders similar establishments so much more seductive: dimmed lights, like eyelids suddenly lowered—because activities carried out in broad daylight are immediately stripped of poetry. Generously lit from one end to the other, its long staircase almost sunny from being located directly underneath the central skylight—so well lit that an art gallery has opened there—it never seemed to offer more than just an opportunity to walk in a covered pedestrian zone, though it is amusing to find the almost identical small range of little businesses there as listed in the inventory of the *Paysan de Paris*'s passage of the Opéra. Today, just like at age fifteen, I cannot find any other rallying point better than the inviting balcony on the top level right in front of the *Beaufreton* Bookstore; overflowing with books, its awkwardly located nooks and crannies still difficult to reach, it remains unchanged. Its next-door neighbor was the famous *Hidalgo Dentaire*, a paradise and almost national museum of gags, games, and devices. Half a century ago, it was already an inexhaustible source of supplies for the modest shows we put on at school and paid for by pooling our meager resources, everything from *fluide glacial*, a liquid that made one's posterior feel icy cold when poured on classroom seats, to sneezing powder and elements needed for making "Algerian bombs" and other *martinikas*, exploding devices which never failed to cause panic during history class.

And yet . . . And yet!—from the balconies with their display windows, from the caryatids and candelabra posts to the bust honoring the donor Louis Pommeraye, *reduced*, no doubt for economical reasons, to the size of a *jivaro* head and perched high above the top of those flights of stairs like the bird of a cuckoo clock (looking down from his shelf with a fixed stare in his eyes, reminiscent of the oracular pose of Edgar Poe's raven)—there is no other image of the city etched in my memory as sharply, or with such photographic precision, as this one. There is still something theatrical about these balconies, this vast expanse of skylights, these busts, flights of stairs, and statues brandishing candelabras. But unlike the theater descending into the streets to enliven them, here it is street life, the unglamorous life of small businesses and boutiques trying timidly to dignify and dramatize itself, and which —under the watchful eye of Louis Pommeraye, paternal muse and incongruous director of this merchants' gallery—has taken off in an attempt to reach new heights, leave the courtyard and garden behind, and scale the walls in the direction of the rue Crébillon.

The west side of the Graslin hill, which descends gently via the rue Copernic, rue de Gigant, and rue Voltaire toward the imprisoned river Chézine, is the very image of a return to bourgeois calm after the animation of the theater district; slightly gloomy and obsessed with exercising its prerogatives, it is a neighborhood typical of many others in the inner suburbs: the waves of silence released by the Protestant church on the place de l'Edit de Nantes, the calm sedative irradiated by the Museum of Natural History seem to dedicate this neighborhood to a vegetative life, protected by an imaginary wall around the secret requirements of the inhabitants' innermost convictions and desires (there remains an unmistakably rigorist, Calvinist touch in the city's physiognomy, almost as famous as Basel's hedonism during the carnival). Moreover, contrary to the Graslin neighborhood where I imagined Eros, Euterpe, and Terpsichore holding each other's hands tenderly while frolicking around a forbidden fruit, I remember this area as being linked to culture, or rather to official culture, by virtue of its most lifeless, most lugubrious rituals. Every time I pass in front of Francine Vasse Hall (then called Colbert Hall; the lady was still alive), I still feel nauseated; a place where, once a year, we had to walk in pairs in order to attend some provincial production of the *Bourgeois Gentilhomme* or *Les Femmes Savantes*, performances that sounded like the tedious, buzzing noises of secular vespers. These spectacles left me with a lifelong disgust for Molière, a radical intolerance I never outgrew.

Only the Dobrée Museum and the Museum of Natural History redeem this complex of austere streets; one adds a touch of fantasy, the other an unexpected smile. I must admit that museums of paintings hardly interest me; but I have always been attracted by museums of natural history, a taste acquired here in Nantes. To enter such an establishment is like finding a cool oasis during the dog days of summer; I love to walk through them slowly, without stopping, like taking a leisurely stroll down an avenue shaded by linden trees. Passing through the rooms, it seems as if all the creatures of the earth have lined up on both sides to watch me, and that these representatives of Noah's ark do not perceive my visit as an effort on my part to proceed with a pedantic, detailed inspection, but rather—since I have been fortunate throughout my life to partake of so many earthly delights—as a polite after-dinner visit. Like the passage Pommeraye, the Dobrée Museum constitutes a monumental bequest to the city by one of those eccentric personalities who are men of substance as well as oddballs; Jules Verne, who populated his novels with them, must have known quite a few of them in nineteenth-century Nantes. Morvan le Besque, my lycée comrade who knew his city well, had more than once tried to persuade me that the indefinable architecture of this building (flanked by a tower—but, alas, hardly ever visited—which was destined to serve as an overnight stop for migrating birds who experienced difficulties while in transit) was definitely an example of the "old Irish" style. Also, according to him (he quite loved to tell stories), in spite of having left detailed instructions of how he was to be accompanied to his last resting place by priests of all denominations, Dobrée's final exit had been a fiasco: on the day of his burial, the coffin slipped out of the pallbearers' hands and careened down the stairway, accelerating its slide from step to step before it finally crashed at the feet of lamas, dervishes, shamans, and other exotica surprised to have been summoned from the ends of the earth to witness an excess of speed.

I am rather fond of this Mr. Dobrée, lunatic and friend of the Enlightenment given to such eclectic preoccupations concerning man's last moments on earth. I don't know why, but I see in him a debonair, already mellowed offspring of those sentimental, music-loving, sadistic slave traders who populated Nantes two centuries ago, and who, thinking even further back, bring to mind an image of Gilles de Rais, burned at the stake in the fifteenth century in the plain of Mauves, in front of a large crowd of people bent in prayer. According to chronicles of the times, people passing through the streets on their way to the execution had been followed by a horde of slaves larger than that of a Roman senator; a noisy horde sounding like exotic birds in an aviary, and so loud that it disturbed the citizens' sleep. On official holidays, gold pieces would be reddened in fire, and then thrown into the crowd for the pleasure of watching the rabble burn their fingers; mores which, a short time later, did not prevent those members of the ruling class from presiding at the philosopher-instigated planting of Liberty Trees while crying "buckets of tears." Strange times indeed, whose flavor is captured by the euphoric, slightly disturbing tone of Sade's *La Philosophie dans le Boudoir*. The nineteenth century, although much richer, did not inherit this ability to incorporate elements of sheer recklessness in any of its artistic expressions of detachment.

No matter what posterity thinks of its donor, I have very fond memories of how pleasant it was to linger in the rooms of the Dobrée Museum on summer afternoons. The light, colored a soft green by the foliage outside, which filters through the latticework of Georgian windows carved of white stone and illuminates the glass-enclosed display cases, is the same as the one falling from the high windows in Vermeer's paintings, impossible to trace to a central source or a single beam, a light gently diffused in rooms like a luminous gas. The feeling of familiar warmth exuded by a house once lived in, which characterizes two of

Paris's museums—the small, charming Musée Hébert, and even the Musée Gustave Moreau (though its contents remind one of a haunted mansion)—turns into a slightly different kind of experience in the Dobrée Museum: one has the impression of having entered the abode of a tireless collector whose frantic pace of acquisitions deprived him progressively of his living space, but in an almost natural process and without drama. Not really a museum, but a house slowly colonized by the animals and props one sees depicted in hunting scenes, who have left the sanctuary of their frames and installed themselves among the furniture; squatters at first, but who eventually became legitimate residents. As such, it joins the other museums dear to my heart, those expropriated human shells which are almost always the offspring of a cohabitation, of a devouring passion, and of chance.

Farther west, the place Canclaux, place Mellinet, and place Zola no longer possess any character or originality, just like the long, straight avenues which connect them; all those loose agglomerations of houses which no longer belong to the inner city but are not yet part of the suburbs look somewhat deserted—they do not bear the mark of the city's genius. I quite often walked along these avenues when I taught at the annex of

Chantenay in 1936: monotonous, gray trenches of stone lined by listless trees where the distances seem endless, where the lonely wanderer sees from afar some forlorn pedestrian slowly advancing toward him like debris floating down the river. I was also never attracted to the rue du Calvaire, one of the city's big commercial arteries north of the Graslin neighborhood, a street that manages to look both lively and gloomy because all the people hurrying by carry parcels or shopping bags; a parade of passersby momentarily depersonalized by a functional servitude. Completely flattened in 1943 during an American bombing raid, it was rebuilt on a larger scale and has since then become even *busier*, its population more anonymous, and almost as attractive to stroll as one of the métro's hallways at rush hour. In the nineteen-twenties, lateral streets led from the rue du Calvaire to two half-deserted squares, each of them dominated by a particular microclimate: in the northwest, the place du Palais de la Justice, just as vacant under its dwarf trees as an empty dry dock, stretching from the peremptory line of columns to the sleepy café across the square; in the northeast, the place Bretagne, unrecognizable today, nothing but concrete blocks and glass walls at the foot of its skyscraper. In my mind, it is only on this side of town, around the place Bretagne, the rue Budapest, and the renovated rue Marchix, that the war made a gash in the urban fabric which has never healed: a fragment of Stuttgart or Dresden *redivivus* inserted itself in a living substance just like a prosthesis; useful, but never an integral part of it. Ever since that time, a special ambiance has disappeared from the city and caused a change in its physiognomy—an ambiance of rural life which had until then managed to survive in the midst of a city whose fairs attracted all the regional products, well known also for its grain and cattle markets, an ambiance that cast a unique light on the ancient buildings surrounding the place de Bretagne and added special meaning to the rhythm of seasons coloring the other side of Nantes, the farmlands opposite the great estuary. This happened at the same time the last old-fashioned shelters of the former toll stations along major roads leading into the city's center were finally dismantled (a project already started before the war), stations where a small tax used to be levied on Nantes' daily commerce with the outlying communities that produced its milk, fruit, and vegetables. These wounds inflicted by the war, and, even more so, the gray, soulless monotony of the new blocks of concrete and glass put up to compensate for the loss have ripped holes in my memory; nevertheless, considering the range of ravages inflicted, they take up a surprisingly small space in the mindset of the wanderer: in the shadow cast by the Tour de Bretagne, the quai des Tanneurs remains unchanged since the days tanners went about their work along its sleeping waters, a time when there were still wash-sheds afloat on the Erdre.

# WHOEVER TRAVELS BACK

in memory to a city he has visited, either as a tourist or as a *pilgrim of the arts*, usually clings to some landmarks as clearly distinguishable from the mass of buildings as are lighthouses for a sailor approaching a port, and almost all of these landmarks are monuments. It is rather peculiar that one tends to concentrate—through a process less natural than what it appears to be—not just the character, but almost the very essence of a city in a few buildings generally considered emblematic, without thinking that such a representation will not only prevent us from getting a feeling for the town's configuration and density of population, but also subtract from its global and familiar presence the power to exalt, create attachments, and make us dream, since our awareness becomes fixed on only a few susceptible points. Taken to extremes, this type of exclusivity—exacerbated and rendered systematic by the growing popularity of guidebooks—can render a town classified as a "city of the arts" almost lifeless for the visitor. The tourist who spends two days in Venice to "see the town" will not have the slightest idea of the spontaneous, charming, and rather simple everyday life just waiting to be discovered along the *calli*, the *rii*, or on the little paved squares. At times like the present, when tourists are being conditioned in advance by the media to see the architectural *musts* in the city they plan to visit, one sometimes wonders about alternative approaches that are more functional, more natural, and less superstitious; for example, not to visit cathedrals unless one plans to attend Mass, or to visit historic houses only if friends lived there, and—since we are talking about Venice—not to cross the Bridge of Sighs unless one's lodgings happen to be in the prison adjacent to the Palace of the Dukes, or cross it only to prolong a state of mind after having reread, once again, Casanova's *Memoirs*.

It is rather in that second way, unstructured and more spontaneous, that Nantes revealed itself to me. Given my situation and living arrangements at that time, I was unable to freely explore and familiarize myself with the city; nor was I simply a visitor. At age eleven, I had not the slightest idea of what monuments, remarkable or not, the city had to offer. No one ever mentioned them to me, and I had read nothing about them. "Cultural" and pedagogical sightseeing walks in the city were not part of the curriculum at that time; I was unaware of the gap which, in today's mind-set, artificially isolates certain edifices from a city's mass of buildings, edifices that speak of its past, which are sanctioned as part of its heritage and revered as masterpieces. I would sally forth blindly, without any preconceived notions, walk the streets of a city that was not separated into various categories, labeled, or catalogued, take in my fill of its masses of stone in all their various forms, its shafts of light, its waterways, its shadowy trenches formed by narrow streets, as one would drink in a landscape, without the least concern for arranging its elements by some order of excellence so as to pay my respects according to rank. I went only where I was supposed to go, or where I was admitted (without having to pay an entrance fee, of course), which is to say almost nowhere. That is why I did not visit the cathedral to see the tomb of François II until I was twenty-five, and never set foot in Nantes' Château, admired by Henri IV (I don't know if I should be embarrassed to admit such indifference when speaking of a building distinguished by *three stars* in the guidebook). Like many habits acquired in that city which shaped me, good or bad, the lack of cultural curiosity and repugnance to visit "artistic monuments and objects" has stayed with me ever since. There have been times when I made concessions, either because of a bad conscience or to conform to usage; I am not sure of ever having profited from going against my better judgment, and from having deviated from my long-established, customary ways of walking in a city as one would in a garden. The sight of simple dwellings providing shelter, like those rows of houses built on piles at Port-Communeau or on Feydeau Island which resemble a petrified seawall, touches me far more than a register of balconies with intricate ironwork, sculptured masks, and pilasters which decorate the old town houses on the rue Kervégan; I also prefer the odor, patina, and texture of a city's surfaces to the architectural jewels which are her pride and joy, often so isolated from their common substance that they sometimes give the impression of being detachable.

Besides, few cities communicate as strongly as Nantes the feeling that there is just a minor difference between buildings of distinction and the run-of-the-mill façades stretching like friezes along the streets. Because of a lack of architectural originality and the unattractive materials used in their construction, most of its churches resemble those found in the nearby countryside of the *pays* Nantais, many of them rebuilt in the last century without the least thought given to style. The unwieldy mansions constructed in the eighteenth century by the slave traders have been gradually abandoned by their occupants, or parsimoniously subdivided like the old Louis XIII town houses in Richelieu; leaning like the Tower of Pisa, decrepit, their paint peeling like that of Venetian palaces on their piles, they are sinking into the dull anonymity of general decay. I always had the impression that whatever was not regularly kept up, refreshed, or renewed by daily life deteriorated more rapidly here than elsewhere.

There are some monuments in Nantes which to me are definitely more repulsive than others; they probably caused that boredom I feel in front of certain forms of architecture. Thus the big-bellied, melon-shaped dome slumped heavily atop the church of Notre Dame du Bon Port, exiled a few steps away from the quai de la Fosse to the corner of a small square shaped like a half-moon, a square unbearably hot in the summertime, had for years inspired me with disgust for all domed churches, a disgust neither St. Peter's in Rome nor St. Mary of the Flowers in Florence could dispel. Only the dome of the Invalides, which I passed with the utmost indifference during the twenty-five years I lived in its neighborhood, could finally cure me— when, one day, in a split second, I became aware of its exceptional power of seduction. But, in reality, the absence of architectural beauties to be saluted brought the town immediately closer to me in an almost sensual way: zones and parts of a body dear to someone are unrelated to the canons of aesthetic beauty. Even today, I don't think of the city as of a town dotted by famous sites but populated by places where I like to be, either physically or mentally (feelings so strong that they tend to confuse the past with the present whenever I think of Nantes). Thus the minuscule train station of la Bourse, looking so provincial and sleepy underneath the stand of trees sheltering it in the summertime, but where express trains and tramways alike stopped to discharge off a dribble of three or four passengers, a building that disappeared long ago together with its tracks, continues to shade in my memories, as clearly as ever, the beginning of the quai de la Fosse. And at Port-Communeau, still bathed by the Erdre though a tunnel now swallows the river at the edge of the square, phantom-like wash-sheds still float like emblems on its waters, while the traffic above perfectly blends the stream of cars with horse-drawn carts à la Breughel and the staccato noise of wooden clogs on cobblestones.

Nevertheless, the Nantes of today has been marred by a fracture that broke up the strong cohesion of its dense urban agglomeration, once a solid mass hardly cracked by the narrow slits of its streets which for me represented the true image of a city, not just an ensemble of monuments. This fracture came about when two of the Loire's eddies that surrounded Feydeau Island fell victim to a landfill; this gap, like a badly healed scar, has remained an open space which city life has not been able to absorb and completely integrate. More than once, I have been awed by the temporary craters and trenches when one of a city's old, dilapidated neighborhoods is razed, or by the more durable impressions of war ruins left intact, or by historical ruins respectfully preserved, spared from the assaults of urbanization. During a number of years, prior to the construction of the Pompidou Museum, I loved to wander up to the Beaubourg Plateau at nightfall when it was swept clean of rubbish and debris, hemmed in at a distance by the obscure mass of buildings still left standing, whose supporting walls showed the scars of ghost-like fire escapes, buildings illuminated at street level by a string of pale lights spaced far apart: this was the only site in Paris over which moonlight would spread like a damask tablecloth, smooth and evenly like across a clearing in the forest. And, when I visited Rome late in life, I found myself immediately mildly attracted by the Forum, a building site cluttered with construction materials. I was struck by their poor quality and the liberal use of architectural *plywood*; to someone unfamiliar with the historical background, it must look like a flea market offering historical debris rather than an ensemble of noble stones and fragments collected from Delphi or Machu Picchu. What never failed to attract me during my promenades were the fallow lands, the rock-scattered goat pastures of Mount Palatin, totally unexpected sights where windswept weeds grew right in the heart of the city, or the immense, grass-covered, empty cradle of the Circus Maximus, stretched out between rows of houses like an abandoned racetrack, protected against construction projects by some municipal taboo. I'll never tire of exploring these empty lots, enclaves in the midst of urban developments which survive against all odds, solitary sites where the wind blows freely, restored to the wilderness and the flora they sustain, land where it seems as if salt had been strewn on the ground, as was the custom in cities like Carthage to assure that nothing would ever grow again; the air one breathes at these sites, while walking on land swept clean by the wind of memories' suffocating alluvial deposits, has more than elsewhere a taste of freedom.

In Nantes, however, the central area created by the

landfills which replaced the river's two eddies is proportionally too large; instead of gaining a distinctive space, the city lost its equilibrium. In almost all other neighborhoods, it is easy to superimpose in one's mind an image of the old over the present picture, but not here. The former image has become blurred; today, it would be impossible for me to point out, even approximately, the location where the bridges once stood. One of the reasons is that an intermediary stage has wedged itself between then and now, an odd, but almost dominant time frame: the actual time of the landfill. I still keep inside of me that disturbing photograph of a river of sand, the wandering riverbed of a Saharan *wadi* run dry, spilling torrential masses of sand between two rows of houses which seemed to have hastily retreated at the moment of an unexpected hydrographical cataclysm—while lines of cars in search of a parking space slowly crawl along the treacherous, rocky thalweg, advancing with the same precautionary movements as horse-drawn wagons testing the solidity of a frozen river. That African vision has since been replaced by a hybrid zone which traffic is unable to take advantage of efficiently; half highway, half public gardens, a complicated system of lanes where cars must navigate alongside strips of lawn, shrubbery, and stretches of asphalt: more of an intersection than a straight thoroughfare, just as inconvenient as the switchyard of a train station, and more difficult to cross than the bridges over the former eddies. In Paris, the line of the ancient fortifications I still knew has been completely obliterated by the furious life which has precipitated itself in the open trenches of its beltways; but in Nantes, this urban renewal project conceived on too large a scale leaves me with an awkward feeling, because up to now it has been unable to blend into that harmony which the organic growth of an agglomeration generates almost automatically. What could have been done with that virgin space created suddenly in the heart of the city? Every time I venture forth on this newly created obstacle course which no pedestrian crosses without risk or fatigue, I can only conclude that the renovation after the war has been a failure, a blow to the genius of a city so rich in interesting sites. It seems to me that mirages should have materialized and taken shape from that sea of sand whose unexpected transgression into the inner city I was able to witness for a moment; but neither Dobrée nor Pommeraye—not to mention Jules Verne—has found a successor here.

There certainly is much to be said—and comments aplenty have been made—about a tourist's indifference toward a city that offers only second-rate cultural attractions (although the museum owns some of the most striking paintings of de La Tour, and one of the most sumptuous portraits by Ingres imaginable). How can one explain that this city—not immense in size, not pleasing to the eye, where three quarters of the buildings are occupied by governmental agencies, whose original foundation on the Loire has been artificially altered by landfills, a "regional metropolis" without a definite sphere of influence, situated at the mouth of a river blocking itself—projects so strongly the image of a "great city"* whereas others, just as large, better laid out and more beautiful, give the impression of being populated by peasants, in town for just a day to shop and run their errands? Perhaps because Nantes is more imperiously centered around herself, less dependent on her territorial and fluvial roots—perhaps because such an impression reinforces her image of being self-sufficient, capable of sustaining an autonomous, exclusive city life, something felt instinctively by the visitor, and which inspires a desire to live there rather than just visit, to immerge himself, and to become privy to the secret

*"I had not yet taken more than twenty steps while following the man who carried my suitcase when I realized that I was in a great city" (Stendhal, *Souvenirs of a Tourist*).

117

of a uniqueness impossible to comprehend. Keeping everything in its proper perspective, it is odd that while re-visiting Nantes, I sometimes think of another city, Madrid; they have nothing in common except a certain, superb *nonchalance* about monuments. It is the small, shadowy bars, the little narrow streets with their tall façades all around the rue Crébillon which at summertime create that same atmosphere of refreshingly cool, subterranean shelters I found in the canyon of lateral streets along the *Gran Vía* in Madrid. Places that leave you with a feeling that native life with all its customs and rites, difficult to penetrate from without and almost entirely wrapped up in itself, can be perpetuated here just as in a network of grottos, independent and self-sufficient from morning to night and into the next day. It is that special tone, that unique, attractive coloring taken on by everyday life, the product of a long and subtle distillation in which its geography as well as its history must have collaborated, something that could not have been achieved without an alchemic transmutation whose formula remains a secret—which might well constitute the real seduction, the crowning glory of a city.

# LOOKING AT MAPS OF A

city's agglomeration will confirm that the relationship of an estuary port with its river is rarely comparable to the axis of a geometric figure. Rouen and Bordeaux are not really firmly planted on the two riverbanks from which they rose. Bordeaux's crescent,
whose inside curb follows the winding river, holds
within its arc an urban agglomeration shaped like a
shriveled up kidney, a sparsely developed annex located on the right bank which only stretches its tenta-
cles along a starburst of roads. In Rouen, the river
draws a rigid line of separation between the heart of
the city and the outlying communities of Sotteville, a
line cutting off the beautiful neighborhoods along the
north bank from the south side's industrial *commons*,
warehouses, polluting factories, and workers' suburbs. During the war, after I had stepped off the bus at
the terminal on the Seine's south side and walked
across the bridge, or while waiting, in the little smoke-
filled station on the left bank in Saint-Sever, for that
strange night train to Caen (a freight train with only
one, unlit passenger car), I had the acute feeling of either leaving or re-entering a zone whose buildings are

reserved for servants' entries and staircases. Nantes is no exception to the rule; worse yet, as already mentioned, the city in the past has never completely succeeded in leaving its mark on the south bank—at least not up to the time when, approximately thirty years ago, suburban housing developments began to take over neighboring rural zones. Furthermore, while crossing the city from upstream to downstream, there is a complete change of character in the river and its embankments; the Loire, which used to reach up into the city's heart with two of its northern eddies, found itself excluded, rejected by the landfills. There has not been, there could never be a divorce, but in a certain sense—a very sensitive point for those who knew Nantes "before" and "after"—a legal separation has taken place between the city and its river.

Sixty years ago, one would approach Nantes upstream from across vast stretches of vacant land: a continuation of the submersible *prées* (but much larger in size), those periodically inundated, unfenced terrains in the Loire Valley west of Angers, such as the prée of Anetz, used as a makeshift airfield by a Messerschmitt squadron in 1944; or the one at Rochefort where, on the rocky elevation point next to the road, a small monument commemorates several of aviation's pioneers, René Gasnier among them, who made his debut here during the first years of the century. There once flowed an untouched, pristine river with nary a fisherman or a boat among these vast, empty grasslands fond of veiling themselves in wintry mists; land unprotected by dikes which sometimes would break up into islands, like the Héron or Beaulieu Island (in those times still quite deserted on the east side). The river, winding its way through those half-drowned lowlands surrounding Nantes, seemed to prepare itself for disappearing, uninhabited, into the marshy arms of some New Zeeland. I loved to see that Dutch aspect of the fluvial approach to the city on days when our Thursday school promenade took us into that area, where herds of cattle rested blissfully in a luxurious solitude. It was on that plain, the prairie des Mauves, one afternoon while I had stretched out on the tall grass and was looking at the Loire flowing by flush with the meadows, that I suddenly had an odd, quietist illumination: a vague feeling that location was irrelevant, that it was perfectly enjoyable and satisfactory to be here or elsewhere, that there was an immediate connection between all possible sites and all moments, and that space and time were only universal modes of confluence. If I compare this intoxicating and completely satisfying sensation of passivity to an illumination, it's because it lasted a relatively long time, at least two or three hours, and then gradually faded away. Never again was I overcome by such an immediate, unassailable feeling of happiness; I don't consider it significant, but every time I find myself surrounded by vast areas of grassland the memory comes back to me—be it the grassy knolls of the Ossenisse Peninsula in Holland's Flanders region, a true pastoral enclave I stumbled upon in May of 1940 at daybreak after marching all night—or the sloping pastures of the Auvergne's *planèze* de Salers, dotted here and there by a *buron*, an isolated farmhouse, where herds of cattle graze during the summer months—or the high plains of Aubrac and Cézallier, where one has the impression of walking on a lunar landscape covered with grass growing wild. A text I wrote thirty years ago, *La sieste en Flandre hollandaise*, is marked by that sentimental attachment which has anchored itself inside of me. I am completely charmed by that feeling of placid tranquility evoked by certain Dutch paintings, where a little town's roofs and steeples are outlined against the horizon of an open meadow. Even Vermeer's *View of Delft*, though more closely framed, emphasizes in an almost abstract manner the bareness of the empty fields on the outskirts of the city, since there are no trees, and no bushes (a French painter

122

of the seventeenth century would have judged this omission unacceptable). The implied safety of such an approach to the city—cleared of every screening device or hindrance—suggested by this view is certainly an element of its seduction: a city thus exposed by the nudity of its surrounding countryside seems to tell us, "This is how I am—for all to see, and with nothing to hide—no matter from what side I am approached, this is how I am." This carefree approach from alongside the river right into the heart of Nantes, seen from afar as the city rises from the tall grass, its skyline dominated by the heavy, rounded mass of the single-nave cathedral, is no longer possible today; it survives only as a stubborn projection of memory. It disappeared because the southeast sector of the city offered the most suitable, flat terrains for postwar urbanization; this area has seen the most complete metamorphosis during the last forty years. A landscape of towers, housing blocks, and *high-rises* has sprung up in front of the old neighborhoods like a capricious nursery of concrete products, where certain more vigorous varieties seem to have shot up in order to gain more light, whereas others, close to the earth, spread out like ground cover; on this side of the city, the Loire no longer resembles a Dutch river between flat banks but has transformed itself into some kind of Texan *río*, hemmed in by skyscrapers. The streets separating these blocks, these towers and high-rise buildings have, oddly enough, been placed under the patronage of the Latin and Greek classics: rue Virgile, rue Senèque, rue Tite-Live, rue Plutarque; or, even more bizarre, were named after both right- and left-wing politicians exhumed from among the most faded glories of the Third Republic: rues Louis Marin, Alexandre Millerand, Gaston Doumergue, André Tardieu, Anatole de Monzie, François-Albert, René Viviani, Léon Bérard, Louis Barthou (the present toponymy of Nantes' streets consists of a series of superimposed layers, all of them rich in fossils of a problematic nature, a stratification which will fill some local scholar's leisure time in future times). The Americanization of the landscape becomes even more evident when one looks at the river embankments; areas where nothing is expected to wash ashore remain at the mercy of the river's whims. Along the banks of the Loire's Madeleine arm, it looks as if the thrust of the current had carved holes in the granite re-enforcements of the quay constructed some time ago. The bank of the island plunges toward the Loire in an empty, grass-covered stretch of land, studded by vertical, masonry re-enforced blocks of stone like a playful display of a geological outcrop; here and there, large fragments of pavement seem to have been dislocated and strewn about by a capricious flood. All along the Loire's Pirmil arm, where the low tide exposes mud formations reaching as high as the top of the embankments, the onslaught of currents caused by abusive dredging reveals the roots and stumps of many willows and alder trees, an unsightly view that keeps the fringes of Rezé's suburbs from creeping closer—until now nowhere in sight all along the southern bank. At ebb tide, a gray, violent river rushes through the abandoned channels; at this point, only two hundred meters away from the harbor, one cannot imagine that it will suddenly carry ships. The riverbanks here carry all the marks of a superficial, stop-gap treatment, of maintenance done on the spur of the moment to remedy only the most pressing needs; not only has the Loire been banished from the city's center, but it seems to have been treated like a polluting, cumbersome nuisance, like one of those beltway projects carried out on unoccupied suburban land at an appropriate distance so as not to inconvenience the city, where the vegetation has not had time enough to cover the scars and rock piles.

Two bridges almost continuously clogged with traffic connect Nantes with a south bank which has become unrecognizable on the west side of Saint-Sébas-

125

tien: the meadows of the river Sèvre and their willows have been replaced by a landscape which is neither part of the city nor part of any suburb; seen from a vantage point like the belvedere of Sainte-Anne, it looks more like the abandoned project of a garden-city, where a plethora of plants growing wild has once again taken over, including a shaggy ground cover that spreads among the rubble. In the middle of that not too densely settled region, in the township of Rezé, rises Le Corbusier's *Cité radieuse*; a massive housing complex clouded by smoke, it bears little resemblance to a "residence" but looks more like a replica of the central power plant near Chéviré, located by mistake in a residential area. Here is where that particular ugliness of recent urbanization around the cities' beltways really comes to light: just like the nucleus of an exploded star, the heart of the old towns (distinguished by the gray and blue, or gray and pink-tiled roofs of their houses), closed like a fist around their narrow streets, has become progressively pulverized, lost in a confusing growth of new housing developments which destroys all plant life, an assault on land reminiscent of craters left by exploding bombs. With the proliferation of single residences along the periphery, we can see quite clearly that the concept of a city is fading away to be replaced by the image of a vaguely human, cancerous densification, which continues to spread its metastases and ganglions far and wide. Entire regions of the former countryside now resemble a chaos of urban elements mixed together thoughtlessly, with patches of greenery thrown in here and there, places where everything remains in a stage of badly blended emulsion, without the possibility of arriving at any clear stratification or separation.

But enough of those ecological ruminations.

The estuary announces itself in the middle of the city by an abrupt deepening of the Loire's Madeleine arm—La Fosse—which, immediately after the new bridge Anne de Bretagne, becomes the harbor. It seems to me that in times past it was better connected with city life, and much easier for pedestrians to reach. After crossing the railroad tracks, a walk on the uneven paving stones of the quai de la Fosse between the tracks of mobile cranes was like an exotic, aromatic excursion, a voyage of discovery among rare woods from the colonies, clusters of bananas, sacks of coffee, sugar, and cocoa. All of that former open-air market is now hidden behind sheet metal walls, inside a profusion of hangars and warehouses that abut and cut off access to the river, or in the Babylonian, massive twin concrete blocks of the Magasins Généraux des Salorges which frame, and also span, the quai Ernest Renaud via a three-lane, covered overpass. On the city side, the almost continuous line of ancient town houses along the quai de la Fosse—victims of neglect and disrepair, covered with the patina of old age, the caryatids of their balconies blackened by industrial smoke—has been disrupted, eviscerated by new roads that provide easier access to the harbor but render it almost unrecognizable.

Even though I feel quite upset and disoriented by its present appearance, it is surprising that in my personal repertory of images the harbor of Nantes has never been as important as the eminent role it plays in folklore and popular songs. For me, the real harbor was Saint-Nazaire—because it opened directly on the sea, because the biggest ships were launched from there, and because it was the port of registry of the great maritime queens, the red, white, and black transatlantic ocean liners of the Antilles and Central America line. We usually stopped there every summer on our way to the beach: this was the real gateway to the ocean, where sea breezes perpetually rippled the puddles on the boulevard Ville-ès-Martin, and where the wreck of the *Champagne*, which had sunk at the entry of the channel, had been for many years a landmark on the horizon, the adventurous symbol of both the war and the hazards of the sea. Compared with Saint-Nazaire,

Nantes was definitely a lesser harbor; cramped in its estuary, without bays, without ocean liners, without real letters of nautical nobility, it seemed to me like the storage area of a great department store or a transport company's service yard, resigned to suffer without fanfare through the peak hours of heavy traffic, far behind shiny display windows. But my feelings about the harbor's neighborhood—a mélange of repulsion and awe because of its slightly sordid prestige—were inspired primarily by the very dense gathering of brothels reminiscent of images in Villon's poems, which spread over the narrow, parallel streets right behind the quai de la Fosse. At night, from a short distance, one could see the lanterns with their enormous numbers—Maldoror's true "flags of vice"—light up the clefts of the narrow alleys sloping down to the quai, shiny like the flamboyant silk-lined slashes of a doublet, as well as the garish green and red colors of the signs which covered the walls like frescoes. I only had a farfetched, totally abstract idea of the purpose served by these places set back from the road. The quai de la Fosse itself had already a bad reputation; I never would have dared to climb those alleys while passing through there just to satisfy my curiosity, since they provided access to something which from the very start had imposed itself like a *taboo*, although no one ever spoke to me about it. What struck me the most right from the start was the mixture of secrecy and crass exhibitionism, the brutal, and at the same time humiliating, abject provocation that emanated from that forbidden zone which, in my ignorance, I guessed to be one of the truly inflammatory points of the city. It is well known that the rapport between adolescence and eroticism has completely changed during the last half century. Quite a few years ago, a director active during the period between the two world wars showed me a script—never realized—which attempted to resurrect, and even celebrate, an image of the brothels of the Belle Epoque as havens of domestic refuge, an image he had obviously formed on his own. While reading it, I suddenly became aware of the gap which separated me from his generation in this respect, just like that other gap which separates me from the next one. The drama-free integration of prostitution into a *well-ordered*, bourgeois family life, the thought of a substitution acceptable to such an existence are ideas which already color the novellas of Maupassant; they reappeared, rather inanely idealized, in the script I was reading, but remained for me—perhaps because I considered them in the ancient light of the quays in Nantes—totally incomprehensible. And, since I quoted Lautréamont, it is precisely that sulfurous, half-supernatural light cast

130

on prostitution throughout the *Chants* which continues to illuminate it for me: the sacred element—in the original sense of the Latin *sacer*: consecrated to evil gods, so familiar already to Baudelaire as well as Mallarmé—continues to blot out all other aspects, and transforms even the most vulgar ones in its black magnetism. Vis-à-vis prostitution, my reaction has never changed since the day I stood stock-still, frozen to the spot on the quai de la Fosse: frightened, dazzled by its sordid aspects, obscurely fascinated.

That entire area behind the harbor, secretly swarming, radiating a venomous power of attraction, where strangers brushed furtively against each other, has undergone a complete metamorphosis. In the last few years, pneumatic drills have continued the destructive efforts begun by wartime bombings, and opened vast breaches in the maze of caves colonized by birds of the night. Sterile, bright daylight now shines from the sloping alleys down on the quai, chases away the phantoms, and restores that area behind La Fosse to the formal banality of *renovated* neighborhoods. Names of presidents of the bar and deputy mayors have replaced the former street names: even the most infamous street, la rue des *Trois-Matelots*, pretty name of an old bawdy song, has been rebaptized; it must have been an effort to exorcise it: I detest street

131

poetry, so dear to Mac Orlan years ago, celebrated in too many bawdy songs extolling sailors' dives and harbor prostitutes, the *Maries-du-port*, and will not shed a tear over the clean-up efforts, probably unavoidable, of a sordid zone. I would only like to say that there was a time when this neighborhood had a haunting influence on adolescents, which is no longer the case. The static electricity created by a city, which feeds that tension particular to city life, is linked to a powerfully contrasting polarization: this polarization, fragile masterpiece created over many centuries, has become the target of the too well intentioned efforts of modern urbanism's collective unconscious.

The single, distinctive feature which rendered Nantes' harbor unique—in my opinion the only advantage over the one at Saint-Nazaire—was a transporter bridge: it granted access to an exclusive club of seaports which included only three other members: Marseille, Rouen, and Rochefort. I never saw the bridge across the Vieux Port; I must have seen the one in Rouen before the war, but cannot remember it. The one in Rochefort, which I crossed once around 1960, was still in operation for a rather long time after the war (I do not know if it still functions today). Located far away from the city, at the end of a sort of side road leading through meadows, it

straddled the low river Charente whose waters re-
ceded at ebb tide low enough to reveal the strip of gray
mud along its riverbanks. I was disappointed to see
this significant feature of a great city so reduced in its
scope, rendered ridiculous by a transfer to the country,
and—since my feelings about hierarchies concerning
such matters had remained acute—I could not help
but consider it a bad joke, like the relocation of Bal-
tard's pavilion to a faraway suburb. Since Nantes'
transporter bridge provided the perfect frame, for me
it was an inseparable part of the harbor's image, like
the Eiffel Tower set in the perspective of the Champ
de Mars. Even today, I can still visualize its gigantic
structure, feet like those of the colossus at Rhodes,
which linked the quai de la Fosse to the building sites
on Beaulieu Island; I remember how, on working days,
it transported—slowly, majestically—its cargos of
workers clad in blue overalls from one side of the Loire
to the other. I have never crossed it. These great con-
structions of lacy steel always seemed to me essentially decorative. Since they are superbly anti-func-
tional, I can only think of them as being "beautiful
like a fern," Arthur Cravan's assessment of the Eiffel
Tower. The bridge only spanned the vast enclave of
naval yards and factories which occupy the tip of the
island, and was used primarily to transport workers;

but, like the Eiffel Tower, it also attracted suicide candidates. Its silhouette remains linked in my memory to the *jump of the Pole*: an immigrant worker from Poland who had figured that he could earn some money by jumping from the bridge's moving platform before an assembled crowd, something he had successfully done once before, in Rouen. On the day of the event, he climbed up the steel framework in front of thousands of gawkers who had gathered on the quais after being alerted by the press; after dressing in what I believe was a fireproof jumpsuit, he hesitated for a moment before dousing himself with gasoline and then plunged down in a burst of flames, like a Wagnerian finale. He did not reappear. The crowd remained silent for a moment and then hesitantly began to disperse, uncertain whether or not the spectacle was really over.

The transporter bridge, which monopolized the view because of its incongruous stance above the waters, has been demolished; today, one needs to climb up the rock of Sainte-Anne in order to get the best view of the entire harbor and an idea of its character. At the end of the solitary little avenue bordered by sickly looking trees, which leads from the church down to the river—a scenic outpost characterized, like so many others on the city's fringes, by an air of resentful poverty—one can see a small stretch of lawn to the right of the statue of Sainte-Anne blessing the harbor from up high on the rock; on the left side sits the little museum dedicated to Jules Verne, who must have come here quite often to contemplate the river from these heights, and take in the site where it becomes the gateway to the sea, and to adventure. Farther to the left, the city is almost eclipsed behind the line of houses along the quay; only the location of its former islands can be seen beyond the space where the eddies once flowed. Under the Nantes sky, so often covered with clouds, the panorama of the harbor and its river looks like a vast, ponderous symphony in gray, barely nuanced by bluish highlights cast by slate roofs and sheet-metal surfaces, and where—unless a freshly painted cargo ship happens to be docking at the quai des Antilles—even the rare touches of vivid colors which brighten Marquet's harbor paintings are sadly amiss. At first, the blunt tip of Beaulieu Island captures the view, visible in its entirety from this observation point on top of Sainte-Anne's rock: a sort of compressed mass which is part of the harbor, squeezed together by the river's arms as if in the grip of a pair of pliers, without an inch of empty space—a confusing jumble of interweaving, flaking surfaces as familiar as that of roof tiles, sheet-metal hangars, covered slipways, dry docks, cranes, warehouses, stores, mazes of railroad tracks. An active, watchful, and congestive uvula in perpetual motion but which at this point divides only two deserted channels: the Loire's Madeleine arm on the left, where the concrete towers of the new hospital form the *skyline* of a miniature Manhattan at the tip of the former Gloriette Island and, on the right, the Pirmil arm with its cloudy, slimy, yellow waters and rotting riverbanks eaten away by erosion. The view ends at the riverbank of Rezé which lies completely flat against the horizon, a crowded suburban settlement with its housing "blocks" and groves of meager willows and scattered stands of trees, stretched out as far as the eye can see around the central dungeon of the *Cité radieuse*. Downstream, toward the right, where the houses of Sainte-Anne begin to obstruct the view of the riverbank, one can guess the location of Trentemoult, a village nestled on narrow streets around its place des *Filets* and close to its church spire rising straight from the bank along the life-sustaining river. A simple relic now of the arts and crafts and entire folkloric past of a river that was a vital source of food; its fishermen's houses, the *bourrines*, once largely isolated from the hinterland by vast, dew-covered meadows and by a small *boire*, the

136

little gulf of the Seil de Rezé, now look on melan-
cholically while suburban housing developments
draw closer and closer to the edges of the village.

It is a vast, but not beautiful landscape marked everywhere by signs of numbing industrial labor, and by the
traces of brutal contemporary mutations in its habi-
tat; a landscape so unlike those Mediterranean seaports dear to Valéry, with their clear skies, incessant
movement, alive and overflowing with noise, colors,
and odors. Here, all the former human hustle and bustle has withdrawn, cloistered itself in its concrete
caves and sheet-metal cathedrals; one can barely hear
the vague noises rising from the conglomerate of factories along the harbor, enterprises whose ships—
never more than one or two—lie asleep at the quay,
more pretext than legitimate reason for their existence. As a matter of fact, this somewhat languishing
dead end has become the neglected, almost abandoned
backyard of a complex zone of docks fifty kilometers
long that extends all the way down to Saint-Nazaire,
a network that has gained in density, and become
more diverse over the last half century; it breathes life
into the city, but leaves its quays deserted. The harbor,
stretched out along the estuary, hardly has any con-

tact now with the city except in an abstract way, via a railroad extension that leads to a switchyard; and the neighborhood of La Fosse—which, ever since the landfill, is now bordered upstream only by a median strip —reminds those who knew the old Nantes of a line by Tristan Corbière: "... vieille coque, au sec degréé, où vient encor parfois clapoter la mare" (old ship, at dry dock, against which the sea still laps once in a while). Surely not waiting for death, like the ship in the poem; but the flux beating the shores no longer throws flocks of sailors at regular intervals in its narrow streets and alleys, and its avenue, now flooded by cars, no longer leaves room for something which in the olden days could be felt in the innermost chambers of that faraway honeycomb: the breath of the sea.

The river Erdre, now swallowed by the archway of a tunnel at the northern edge of the cours Saint-André, emerges again next to the Lefèbre-Utile factory from the Saint-Felix canal—reminiscent of the discreet exit of a main sewer line rather than a river going underground; its disappearance from the center of Nantes is perhaps even more noticeable than that of the Loire. Its narrow riverbed sheathed by vertical walls of granite, fitted with locks like a canal, once marked the border between medieval Nantes and the Graslin neighborhood: just a small passageway of walled-in water, as placid and inert as a *gracht* in the Netherlands. It is much easier to guess the scar line left by this landfill along the cours des Cinquante Otages than those left by the fill-in of the Loire's former eddies; the adjoining rue de l'Arche Sèche, which runs almost parallel on its right down from the heights of the Graslin neighborhood, a street crossed by the rue de Feltre, the rue des Deux Points, and the Sauvetout Bridge, almost looks like the original riverbed underneath the arches of its bridges. I don't know if Nantes, in order to be at a safe distance from the caprices of its great stream, had chosen the banks of the Erdre rather than those of the Loire for its original site, like Strasbourg which first rose alongside the Ill, or Lyon next to the Saone. The distance would have been minimal; on the other hand, it would be difficult to find two rivers as different in character as these two. Just like the Grandlieu Lake, southwest of the city—a shallow maze of river arms, mudflats and reed beds, which during the wintertime expands into one of France's biggest lakes —the Erdre has witnessed a recent drop in the level of the *pays Nantais*, and seen more and more dikes rise throughout the valleys, a change which has significantly affected the entire run-off and drainage system (everywhere except the Loire). Upstream, above

Nantes, the Erdre is like an Irish river, a body of water with almost no current; it narrows and widens here and there, stretching out to embrace and annex bodies of water such as the vast flooded areas which surprise the wanderer at the foot of the Sucé rock formation. Even through its course through the town, the river regularly grows wider upstream. Just one kilometer before entering the tunnel, it is already wide enough to surround a small island, the île de Versailles. At the Tortière Bridge, and even more so at the Beaujoire Bridge, it is no longer a modest local tributary but a *body of water steadily growing;* by the time it reaches the parc des Expositions, it has grown as wide as the larger one of the Loire's two arms.

That is why all the pleasures associated with calm, mirror-like waters, pleasures the rapid and brutal Loire refuses to Nantes, have taken refuge along that odd, paralyzed river. Of all the rivers I know, it is the Loiret flowing downstream from Olivet which best reminds me of the Erdre's banks: an elongated, miniature lake, where villas on both sides of the water are set amidst lawns, contemplating each other in the intimate, tranquil setting of estates in a private park. However, Nantes' waters are more crowded. Starting at Port-Communeau, the Chinese-like invasion of the river by a flotilla of all kinds of small boats contrasts markedly with the deserted Loire. A little farther away, close to the Versailles Island, are the docks of the two-decked *vaporetti* serving summer tourists. Beyond la Tortière, a great variety of green zones, recreational sites, and public gardens were established along the riverbanks after the war: the *campus* of the new university, an exhibition site, the training fields and sports installations of Nantes' Football Club, a nautical center, and a vast *playground* which reaches as far as the gates of La Chapelle sur Erdre. Someday I shall go again to see the Erdre in its present state, most likely aboard one of those boats for Sunday tourists, because all the roads and streets past the Tortière Bridge are at some distance from the riverbanks. But I fear that its official consecration as a pleasure zone will have spoiled it a little, since administrative approval tends to rob landscapes of their charms. Worse yet, I am afraid that I will never again find that special charm which I felt during my first and only excursion on the Erdre—a feeling of surprise and delight, of having entered a strange, but wondrous realm of protected intimacy. The lycée's administration, not known for spoiling its boarding pupils with organized recreational treats, one day had the fantastic idea—unique in the seven years I spent there—to replace the mortally boring Thursday session of educational cinema

141

(from one o'clock to two o'clock in the afternoon), together with the obligatory promenade that preceded it, by a boat ride—a real excursion lasting an entire afternoon—on the Erdre. We could hardly believe our good fortune when, resigned in advance to suffer some treacherously disguised, supplementary cultural enrichment ordeal, we embarked on the other side of the Versailles Island on a small motorboat hired for the occasion, and installed ourselves on the banks of the upper deck under the open sky. I have very few detailed memories of that excursion; they have drowned in a very strong overall impression of an absolute feast, a calm feast that lasted the whole afternoon, without any specific moments of special excitement or pleasure, and thus even more enjoyable. An enchantment because of the expense incurred in renting the boat, exorbitant in our eyes, which made us wonder if the administration had unexpectedly fallen under a spell and dipped into the till; because of the unexpected treat, a nautical excursion squeezed in on a school day; and because of the wonder of the unknown river which grew larger as we approached its source. What I found again that day, on a larger scale and under quite different conditions, were certain pleasures I experienced as a child while boating on the Evre in Saint-Florent. Pleasures such as that feeling of intimacy, quite similar to the one inspired by a quiet garden path, which takes hold of us while we glide downriver in the heart of a valley where there are no roads, where there is only silence; no one can ever reach the heart, or feel at one with a landscape, unless one has traveled across it on a river, the peaceful effusion of its liquid essence. I believe that Edgar Poe's brilliant intuition in the *Domaine of Arnheim*, where he tried to convey the idea of what might be a masterfully composed landscape, was not to take the visitor on a walk, nor to have him embark on a rowboat or sailboat, but simply let him drift downstream in a skiff, pulled along by the current. No doubt that the silence on our excursion was less than perfect; but, just like in *Le Grand Meaulnes*, the little motorboat only made "a calm noise like that of a machine and water," the banks flowed by without haste, every bend in the river seemed to raise a curtain on an intimate scene, like a half-open door offering a glance inside. The banks of the Erdre are hardly visible: the recent drop of the surface level in that area makes them look at times like a low-lying valley upstream of a dam, in a region that is all but hilly. What I remember is a landscape of reeds growing along the riverbanks, stands of trees crowning the higher regions, solitary country homes looking longingly at their reflections in the mirror-like waters, ready to slide into the river but held back midway by their lawns, an almost luxurious, very sparsely settled, secretive suburban enclave set along the river which jealously guarded its privacy. The end station of the excursion was the roc de Sucé, a rock formation which I see again in my mind as towering directly above the river, crowned by a few pine trees—but I could be wrong, because memory has superimposed the familiar image of the rock of Courossé rising above the Evre. Beyond this point, an unknown Erdre flowed into a real lake, the Plaines de Mazerolles: it was like the threshold of a new country, of a superior headrace, where the horizon abruptly widened; a region where it could be said that the laws of hydrography were only absentmindedly observed.

It is not just the banks of the Erdre which over the last sixty years must have changed a lot, but my image of them as well; probably deformed by now beyond all measure, and as unconnected to the reality of those times as some trivial event of daily life is to our dreams at night, but which a dream can open and make bloom, like a flower. It is simply the atmosphere of dreams which intervened here and took over—a rather rare occurrence—while the film depicting real life kept on rolling. Moreover, I do not in the least pretend to paint the true and faithful portrait of a city which has never allowed me to see the light filter intact through its prism. As I have said, I only want to give an account of its presence inside of me: the only one, of all the cities I have known, which stands independent of verification.

# IT IS MORE DIFFICULT

for me to recall life in Nantes during the nineteen-twenties than to remember the city's former configuration. The movement on its streets—the most unstable and most volatile element of a city's image—escapes me; I believe that this blank in my memory is caused, at least in part, by the fact that this was a period of transition when, for several years, the horse and the motor still shared equally in providing transportation. Big trucks were practically unknown; horse-drawn wagons (including those of breweries which, for several decades, remained the last specimen in service) continued to transport and deliver merchandise; but there must have been the occasional automobile, *coupé*, sedan, or *torpedo* (convertibles then widely used for touring) parked along the sidewalks from time to time. Since the garrisons were still overpopulated in those years immediately following the war, the sight of uniforms was probably much more common than today. *Coiffes*, the white starched headdresses worn by women, dotted the streets like moving constellations: the maids, almost all of them from Brittany, wore them every day. The toll stations at the city gates, the wash-sheds along the river Erdre and at Port-Communeau were still in service. All these vaguely picturesque images did not leave any distinctive imprints on my memory: the only original figure of that era I clearly remember is the lamplighter, a Dickensian, taciturn silhouette, friend of winter's twilight, who emerged from the evening mist armed with a long pole ending in a curious apparatus with double action: a snuffer surmounted by a short igniting rod which topped every gas lamp with a little tongue of flame the instant he touched it. The memory of that silhouette immediately brings to mind the distinctive image of the city's streetlights, somewhere between the feeble illumination found in churches and today's garish neon lights: a yellowish, semi-mystical light, trembling in the winter wind, leaving all the shadowy corners to their secrets, and not always successful in its plight to resist the onslaught of fog rising from the Loire.

Moreover, my contacts with city life were only indirect. Cafés were off limits to me; I only went to the cinema (then considered a form of entertainment far less prestigious than the opera) on those rare days when my family came to Nantes and *rescued me* from boarding school. I hardly knew what the streets looked like, except for what I saw on Sundays. Nantes' rumors and gossip penetrated the lycée as if through a filter, in the most fanciful manner, whispered by the day students during class hours. The city's noises, softened by the distance, came through like a homogenized, almost unintelligible buzz. Nevertheless—or perhaps precisely because of that distance which separated and sorted out what could be perceived—it seems to me that right then and there, I developed for the first time a certain idea of society whose traits nothing could ever completely erase thereafter.

I had arrived from a rural burg beset with its share of tensions during normal times, but where the war, like an enormous vacuum cleaner, had been able to absorb temporarily all the poisons of a social body without great girth. It seems to me that the war of 1914 had been for the villagers who had stayed *at home* an era of spontaneous, profound truce. Its tragic aspect had escaped an eight-year-old child, but not the impression of how it felt to live and breathe in the midst of a small, ordinary society almost perfectly at ease, where no post was disputed, no order of precedence contested, no belief discussed.

My initiation into a more complex social mechanism took place at the lycée (each *département* then had one lycée, and one préfecture, the governor's office), one of a city's institutions best qualified to represent its hierarchies and complexities, but also one of the most bewildering for a child who grew up in the country. Looking back from a distance—and it took a great distance—not all of my memories of it are unhappy memories (but then war also leaves some good ones). All things considered, I acquired a solid basis of knowledge, and I am almost certain that the scholarly results obtained from that hard and brutal machinery have been, for me and my comrades, far superior from what is being achieved today. But the price paid had been high. Looking back after half a century, I am amazed when I think to what extent that institution had retained its Napoleonic flavor (and where the boarding facilities, dear to the Company of Jesus, continued to occupy a central place, although its mission had been reduced from an educational program to a strictly disciplinary system of supervision); amazed also when I realize how its many aspects of aggressiveness are diametrically opposed to the contemporary dream vision bewitching today's *convivial society.* Order, uniformity, hierarchy, and discipline had been the key words. It was a school of hard knocks where one was forever bumping into sharp corners, where every spontaneous movement could leave a bad scar, where almost all the situations were uncomfortable, from the ice-cold dormitory to the parsimonious supply of linens to the fish served at lunch which smelled of ammonia, from the drafty hallways to the appetizer of castor oil before meals (although, it must be said, administered only at the request of one's family); the ordeal of its ingestion comparable to that of the little school-age convicts in *Nicholas Nickleby* forced to swallow sulfur-laced purgatives. An institution that imposed the law of no recourse, where the distance between high and low was almost impossible to transcend: a student's request to speak to the principal would have seemed just as incongruous as a recruit asking to see the general of his division. I have written elsewhere, in an essay on Lautréamont, about the sporadic outbreaks of anarchical impulses resulting from such constraints. But the administration was adamant in its policies: I remember that following some student *pranks* which went too far, about thirty students were suddenly expelled, just as casually as that contingent of Soviet diplomats some years ago. But those times are long gone.

However, in spite of being so carefully shielded from outside influences, the era's climate had penetrated the lycée, and with it an awareness of the movements which agitated *civilian society* (a term coming to mind automatically, because the lycée's ambiance so much resembled that of military barracks). And, from what we were able to guess, such news reinforced the atmosphere permeating that secular cloister in which we lived: the spirit of the times was neither a laissez-faire attitude nor inclined toward permissiveness. Everything I saw around me during the first years of the nineteen-twenties, years still marked by the First World War that had just ended, appeared to be of a timeless stability, set under an unchanging light; only the 1924 elections began to introduce slight changes. Today, while going through my memories in the light of book titles published in Paris during those years, *La fin de Chéri* or *Le bal du Comte Orgel*, I realize that the *Directoire* atmosphere, the loosening of moral standards, the thirst for pleasure of the hedonistic Roaring Twenties had hardly touched the cities in the provinces. They came back to life at a much slower pace; the crowds in the streets were still ostensibly darkened by mourning clothes, and the spirit of the hour was more likely that of *le Bloc National* than the ambiance of *Le Boeuf sur le toit*. Even though the concept of the "sacred union" in university practices had lost its appeal as the massive vote of the teachers in favor of the leftist *Cartel des Gauches* showed in the 1924 elections, some of its aspects had survived the war; seen from a distance, it appeared as if life in the lycées had not changed much since the Second Empire. The Church, although not welcomed with open arms, was not held in check as much as it must have been before the war, during the time of the *Bloc*. Two chaplains, familiar sights one would run across in the school's hallways, directed the *retreats*, prepared pupils for their first Communion, and provided daily religious instruction, classes included in our schedules and taught on the premises. Wimpled nuns were responsible for the maintenance and decoration of the chapel, the laundry, and the infirmary. At the top of this hierarchy was the principal; installed like a visiting bishop in the special stall of a church canon, on a lateral platform elevated by two steps (making it look like an official enthronement), he presided every Sunday during Mass in the chapel. The aftermath of war remained tangible in the special consideration expressed by officials for the *pupilles de la nation*; I remember the war-like deployment at the 1922 secular dedication ceremony of the lycée to Georges Clemenceau, where music played by a military band solemnized the annual distribution of prizes. All these impressions must have left a child with the image of a solidly constructed and smoothly functioning social machinery, of a *consensus* where the army, religion, and civilian society were closely interwoven and cooperated fully (I remember a former corporal eager to continue practicing his skills who, for several years after the war, enriched the liturgy at the church of Saint-Florent on certain religious holidays with some resounding trumpet solos: a melange of genres not without eloquence). When I reread Louis Guilloux's *Le Sang Noir*, and compare his vivid portrayal of Saint-Brieux's lycée in 1917 to my experience in Nantes' lycée five years later, I realize that there must have been a drastic reduction of tension during that interval (the war was already fading away); there had been no fermentation at my school, only a frigid conformity feeding on itself, still running its course. But that society—direct heirs of the carnage—bled dry and without force to extricate itself from ruts dug so deeply, was not suicidal but kept its strong defensive reflexes: an eleven-year-old child living in the microcosm of a lycée discovered the existence of an established order, in the full sense of the word, legitimized by victory, and apparently in place forever.

It was an order imposed and accepted without resistance, though to me it seemed strangely insipid; lifeless and antiquated, it prevailed at an institution where the physical layout of its facilities had its own special eloquence: an eloquence representing a ladder of values—or rather the absence of a ladder—where one's movements were dictated by the *schedule*, and carved in stone. In the center rose the throne-like professorial pulpits, around which all the activities of each working day oriented themselves. At one extremity, directly across the street from the lycée stood the museum; I thought of it as the commons reserved for art, already foreshadowed *intra muros* by the glossy reproductions of Raphael or Andrea del Sarto hanging on the humid walls of the school's hallways. At the other end, in the wing overlooking the rue de Richebourg, the grouping of all the annexes—chapel, laundry, infirmary, bathrooms and showers, projection room—took on a special meaning for someone like me, a child not yet sensitized to the eloquence of such a peculiar arrangement. In my mind, the perfunctory cleansing of the soul on Sunday morning was instinctively counterbalanced by the Thursday morning shower in the bathrooms installed just below the chapel, in a sort of crypt. And so religion and art represented the two sides of the magnified central apse where scholarly Knowledge reigned supreme, the more than eminent and nevertheless devitalized source of any promotion. Although formally approved by administrators and family, religion and art were considered *optional subjects* and thus of marginal interest; worse yet, subjects to be studied on days when school was not in regular session: something not without merit, but only halfheartedly recommended, like homework during vacation.

Looking at Nantes beyond that reduced model of a society as represented by the lycée, I could observe the movement, animation, and comportment of the city's crowds only during the hours when I came into contact with them: my leisure time. That is to say, the Sunday crowds, the only day of the week when they were free to pour into the streets, at a time when weekends in the country were practically unheard of. Although still possible at the beginning of the century, after 1920 laborers and factory workers could no longer be identified by the clothes they wore away from their workplace (a change brought about by the First World War; after the Second World War, the wealthy classes' vestimentary distinctions also disappeared, at least those of the young people). My impressions were limited to representatives of a world where everything related to manual labor was held at a distance, where it existed only as an abstract concept. Rather than a modern city's society, what I saw coming to life in the streets was a strange native population who considered physical work loathsome, something to be banished from view and hidden inside its workshops and servant quarters, while the right to spend hours of leisure in the streets and public places was accorded to a bourgeoisie without distinction, thrifty and, according to today's criteria, almost indigent (for example, I remember the shoes with rundown heels, restitched, mended, forever resoled, deformed by the addition of rubber heels, shoes only bums would dare to wear these days), but keeping its rank distinctions and exploiting its servants, and who pursued, in spite of limited resources, careers which to me seemed exclusively liberal. Yes, after all these years I can only think of that era with astonishment, a time when history programs still put Roman history at the top of the reading list, taught simultaneously with Latin declensions; it seemed as if the society immediately around me had remained relatively unchanged, and followed naturally on the footpath of those ancient civilizations, the first ones we studied in school. Slavery, as practiced in those early centuries, had disappeared only to be replaced by another kind of bondage. What had evolved was a community of citizens both plebeian and privileged who kept its own counsel and fussed about its rights and prerogatives, a society to which I vaguely felt I belonged, and whose distinctive and almost natural attribute was to be exempted from manual labor. Today, living in a society where the social distance between white- and blue-collar workers has become almost negligible, it is difficult to imagine the disgrace then associated with manual labor in the eyes of the lower bourgeoisie, a disgrace tantamount to some horrific, unspeakable taboo, especially by those who found themselves closest to the working class because of their income level and lifestyle: it was the same reflex of segregation at any price which motivates poor whites living among indigenous populations. They would meet, see each other without taking notice: contact was refused. During the nineteen-twenties, when someone in a family of the lower bourgeoisie had sunk as low as to take up manual labor, an unspoken rule kept him apart from relatives and reunions of the tribe; one would see him only at funerals. That partial blindness, selective and imparted, very quickly noticed and interiorized by a lycée student from his milieu and daily contacts, marked everything that I saw and heard— and even more of what I did not perceive—of life in Nantes: an attitude acquired at the end of childhood and at the beginning of adolescence so strong that nothing could ever erase it completely inside of me. Several years later, the riveting text of *Le Rouge et le Noir* helped me to correct my view and adapt another attitude, take a hard, objective look at the generally accepted image of the society I lived in: however, that approach has remained theoretical. The unstructured, erratic timetable of a teacher's schedule rarely necessitates contact with subway crowds or suburban trains during rush hour, and it so happens that all my adult occupations have contributed to keep that barrier in place, self-imposed and nevertheless insurmountable, which very early on in my life marginalized the world of labor. But that's the way it is. Every society, like every landscape, does not come alive until it is observed from a certain viewpoint. I must admit that I shun and mistrust those intellectuals-laborers who, following in the footsteps of Simone Weil, have at times attempted to see a world long familiar to them *anew*, through the "borrowed" eyes of —for example—a worker at a conveyor belt: an experiment that further deforms the old as well as the new viewpoint by adding to it the brutal *trauma* of a sort of second birth (and what a birth! Not the birth of a virgin organism, but a rebirth which must rid itself spontaneously of all its habitually used mechanisms of adjustment).

And so I wandered through the streets of Nantes on my free days, naïvely mingling, elbow to elbow with a crowd dressed in Sunday finery, whose leisure-time habits seemed to me perfectly natural; people served by "maids" sequestered in their kitchens, their meals prepared and needs taken care of by industrious hands toiling far from watchful eyes in a tacitly accepted secrecy. The city seemed to be populated by a vast social *mainstream* where I felt vaguely at home, a mass of people where only one extreme fringe touched wealth, and another, servile labor. There was no tension in the air, and no haste in their movements: the attitude of every passerby appeared to be one of spontaneous benevolence. Unlike today, one would see passersby who did not know each other strike up a conversation at the corner of a sidewalk, prompted by a simple demand for directions, conversations in which the respective civil status, age, profession, residence, family ties, war service, hobbies, and travel would be mentioned and freely discussed. Certainly less lonely, less scattered, and with a more positive attitude than today's crowds, more inclined toward agglutination than refusal of contact; a population of citizens with small private incomes, buoyant, ready to listen, naturally capable of turning into idle, curious onlookers. Nevertheless, twice a year, on Fat Tuesday and midway through Lent, I was able to get a very close look at what the police of the Second Empire had called the *dangerous classes*—an element both disturbing and sordidly attractive, characterized by their manners, songs, and especially their language.

A yellowed postcard, which depicts the 1923 carni

val in Nantes, brings back memories of daring feminine silhouettes that populated the streets for two
days, a more than troubling presence: this impression
of personified feminine provocation, associated with a
paralyzing apprehension, preoccupied me for many
years. On those days, women workers (the vast majority could not afford to buy a complete costume) attached a coarse wig of cropped, curly hair that looked
like sheep fur above a black mask (covering only the
eyes, without the attached little veil). It is cold and often rainy in Nantes during February and March, and
darkness still sets in early; they wore thick, knitted
sweaters that tightly covered the bust and often tied a
scarf around their neck. This wintry gear was completed by short, tight black pants which only covered
the upper part of the thighs, so that the legs, clad in
black silk stockings, appeared to be much longer;
high-heeled, laced-up boots completed the costume.
From their wrist dangled a Basque tambourine; the ob-
sessive, dull noise of its clapper, not cheerful, but mechanical like a harness bell, filled the streets from
morning to night. The suggestion of a risqué state of
semi-undress—rather bizarre since it stopped at the
waist—floated around these strangely wedge-shaped,
incongruous silhouettes, an impression which the
word *déjupée* (un-skirted), a term only encountered in

a text by Klossowski—coined at a time when women had not yet begun to wear trousers—not only accurately describes but perfectly conveys the meaning of such insidious eroticism. Poverty, or perhaps misery, completely blocked out for a day by the anonymity of naked desire, was throwing an intoxicating challenge to the passerby through eyes heavily made up behind black masks. On those days, the ritual lycée walks avoided the crowded streets in the inner city even more than on others: walking single file toward the suburbs, we would meet small, masked groups coming toward us that joined together and streamed toward the city center, abuzz like a beehive. The tinny, feeble noise of tambourines, which every so often burst into a staccato wake-up call, resounded in the already empty outlying avenues like some barbarian percussion ensemble, like a call by the initiated to attend a mystery unknown to me but whose preparatory rituals made my heart beat faster.

Those thick, bosomy torsos on top of legs made to look slender by their black silk stockings, knock-kneed legs with knees slightly pressed together, surprised and intrigued me (I was eleven years old). These insolent, powerfully vulgar silhouettes of instruments of pleasure, who populated the streets for a day and almost outnumbered the feminine population still wearing the gray and black ankle-length skirts of the early nineteen-twenties, have remained for me the first really troubling image of sex appeal, an impulse I was unable to call by name at that time. These encounters have left their mark on me; a certain daring vulgarity in feminine provocation, a hint of coarseness in the expression of desire are impressions which still provoke a reaction on my part.

Aside from the one noteworthy exception provided by the days of carnival folly, street life left me rather indifferent: I was already developing an absence of curiosity (Breton probably would have called it *vibration*) for local mores and the particulars of daily life, something the passing years have only increased. As soon as I find myself in a city, where movement is concentrated in the streets, life always seems to be flowing more or less with the equalizing monotony of a river; it is possible that I am more inclined than others to let a person fade into his surroundings as soon as his individuality starts to become blurred by the presence of other people, a background immediately dehumanized: he has melted into a crowd, become invisible. Nevertheless, I discovered at an early age that there are places where the human element manifests itself in all its glory, though even more subjected to the laws of collectivity: these prestigious sites are stadiums, and it is Nantes which made me discover them.

In the nineteen-twenties, Nantes took up rugby, the first city north of the Loire, I am quite sure, to become interested in that sport; since then, it has acquired a place of honor in professional football. Its top-seeded team, the SNUC (Stade Nantais University Club— white jerseys with a green, white, and red belt), has not left any dazzling memories in the annals of the game; it used to languish down at the bottom of the second division, never reaching—never—the first. But it had a following, fans who braved many a rain, a sterling public both faithful and resigned, unfazed by any set-back, which persevered when there was no hope as long as the *flag-bearing* team held on to the first place in its subdivision. Its matches with the Velo-Sport Nantais (red and blue colors: the *local derby*) and with the R.C. Trignac (the Trignac Rugby Club—"the team of the tall furnaces"), a superb *pack* of brutes hardened by their daily work in the steel foundries, as famous as the gigantic bulls in the plaza of Bilbao, electrified the local crowds. The stadium of the SNUC was a rather primitive facility along the canal Saint-Félix on Glori-ette Island, almost directly across from the present-day Marcel Saupin Stadium. Memory somehow con-fuses its image with that of the wooden stands of the

163

ancient, primitive Parc des Princes, a stadium I remember from long ago, when I first arrived in Paris; it was replaced by the second parc, built of concrete, surrounded by a bicycle racetrack, and then once more rebuilt into the present facilities. Close to the Loire, along the quai Ferdinand Fave, the stadium was enclosed by a wooden fence of medium height: one could watch the games (but not very well) without spending any money, either by squinting through holes in the fence where clever fans had removed all the knots in the wood, or by setting up a stepladder on the sidewalk which provided an unobstructed view of the field from its top rung. On Sunday afternoons, just before starting time, one could see a pack of prospective spectators running in spurts, bent over like mushroom gatherers, trying to find the best holes in the fence, while the privileged few sat placidly enthroned atop their ladders, just as immobile as line fishermen at a contest.

A year or two before my arrival in Nantes, I had started to read *Le Miroir des Sports*, a newspaper already then generously illustrated with photographs. Like all children who played early on in the schools' courtyards, I had an approximate idea of the rather simple rules of playing football, but the *Miroir*'s photographs left me in the dark when it came to rugby. I was immediately fascinated; but then I have always been fascinated by games with convoluted, at first unintelligible rules that seemed to hold a secret. At age fourteen, it had been the abstruse rules governing the movements of chess pieces across a board of sixty-four squares; at age nineteen, in London, I was captivated by the cricket game; at age sixty, in the United States, it would have been baseball if only there had been time. I picked up the rules of a rugby game here and there, on my own, like a child learning to speak. After lunch at my great-aunt's, I hypocritically steered Angèle, who was supposed to *take me for a walk*, toward the quai de l'Ile Gloriette, where my tender age usually helped me quickly to gain access to a rung on someone's ladder. Once installed, it became quite difficult for Angèle—patience personified, walking back and forth on the sidewalk—to pry me loose from my perch before the end of the match. Little by little, the ash-colored fog of the Loire settled on the cold grass; during the third quarter, every time the team came close to the white line, the excitement made my heart beat faster; I would climb down from my ladder and step back on the street, body a little stiff, head in the clouds, feeling like Hemingway—whose quote perfectly expressed my sentiments at that moment— "just as empty, just as changed, and just as sad as after any deep emotion." Those strong emotions remained deep inside of me, there was no one I could share them with. The lycée had a football team: les *Tigres du Lycée de Nantes* (yellow and black striped jerseys). Tigers with rather weak jaws, but who from time to time played an amicable match in Port-Saint-Père, Vallet, Machecoul, or some other capital of a neighboring region; games which, in spite of a chaperone's presence, were simply a pretext for some memorable drinking bouts. But I was apparently the only one interested in rugby. Rugby matches became orgies of solitary enthusiasm; isolated on a ladder, I could not share my enthusiasm by screaming along with the crowd, nor find a sympathetic echo afterwards among friends and classmates while rehashing plays. But before going to sleep at night I would replay, in my mind, all the exciting moments of a game, project them against the dormitory wall as one would scenes from a film. Even today, when I think of that anonymous corner on Gloriette Island, so gray and monotonously suburban (it has since been rebuilt), memories of those orgies transform it into a corner of Mount Olympus's ruins.

The other favorite site of stadium games was an athletic field at the other end of the city, close to the boulevard des Anglais which, as far as I know, is now

called Pascal Laporte Stadium. I went there only two
or three times but cannot recall the circumstances; it
could have been a track and field meeting attended by
representatives from my lycée. I have no clear memories of either the site or the type of competition; it
might have been only a training session rather than a
game. But after that event, every time I attended
a track and field meeting I found again that special charm experienced at Malville: the rare, sporadic
noises, the muffled sound of feet running on a ground
cover of cinders, the hours slowly passing by, the
shimmering grass on which young bodies stretch out
and relax after their efforts, the relaxed, a little disorderly atmosphere of a sports picnic, the sweetness of a
sunny afternoon slowly passing by, hours lazily strung
out by a rosary of exercises taking place without haste.
I had a feeling of having found the true garden of Eden
of sports, flourishing in an ambiance of original inno-
cence, the perfect realm as envisioned by Montherlant
who, during that same time period, extolled the glory
of the stadium (however, in my opinion, much too
closely associated with war games, and the helmet-
wearing Minerva). Sometime later, after accompany-
ing a fellow student to the terrain de la Faisanderie
where he spent the afternoon training, I again found
this enchantment while stretched out on a flowering

meadow, next to the track covered with light-colored cinders, in the shade of Virgilian trees.

I have not said anything about public opinion, or the political situation in Nantes during those years long past. The truth is that I knew nothing, or very little, about it. My first opportunity to comprehend what could excite public passion was the 1924 elections, when the sons of teachers working in public schools took back to the lycée an echo of the jubilation expressed by their parents at the victory of the leftist *Cartel des gauches*: I was astonished. The municipality, most likely activist, cultivated a leftist tradition which the war years must have severely undermined; whatever their sentiments and beliefs, everything was toned down, softened, made palatable until there was not a single sharp edge, and no echoes of conflicts appeared on the walls or in the newspapers: a child my age could spend two or three years in the city without being aware of political parties fighting for power or influence, something unthinkable today. The reason, I imagine, was that the almost universal itch for change shared by today's social bodies was then a thing unknown: the zone of friction, extremely limited and unchanged by the war, only covered the problem of secular versus religious instruction; everything else functioned according to a middle-of-the-road harmony, assuring the cohesion of a "good society" and which, rightist or leftist, had not really changed since the era of Mac Mahon (as mentioned previously, I completely ignored the opinions of people in the working-class neighborhoods of laborers and factory workers). There is one image of that era I remember quite clearly, just as naïve as difficult to forget. One year, the Alumni Association decided—I don't remember on what occasion—to invite a student to its annual banquet. Having received a number of prizes at the end of each school year, I found myself the designated person. On the day of the event, which took place at the banquet facilities of the Salons *Mauduit*, ill at ease, I crossed the city from one end to the other in the company of the principal who wore his customary bowler derby, and, as was his habit, never opened his mouth. After the banquet there was an auction destined to replenish the Association's funds earmarked for benevolent causes. All the lots, without exception, were nonchalantly bid on and bought by a cigar-smoking guest; someone who, in my naïvety, I thought of as a fabulously wealthy American. That aficionado of Havana cigars was the playwright Mouëzy-Eon, completely forgotten today, but whose royalties, at that time, considerably exceeded those of his famous compatriot Jules Verne. He had made his fortune from the production of military-inspired vaudeville plays, especially from the success of his long-running masterpiece *Tire-au-flanc,* an efficaciously dosed combination of stand-up comic's jokes and slightly scatological light comedy. During those years, Parisian theaters showed plays by Cocteau, Tzara, Raymond Roussel; however, the contributions to the coffers of the S.A.C.D. (Société des Auteurs et Compositeurs Dramatiques) clearly indicated the taste of the majority of theater patrons: not quite as low as during the era of General Boulanger, but still feeding at the trough of folklore where, for half a century, the French nation committed to the three-year obligatory military service had looked for comic relief.

In spite of the cautious secularism practiced in the lycée, its pupils had the somewhat confused impression of living on an island assaulted by waves from all sides—though their onslaught could not exactly be called furious—waves unleashed by the inmates of the many religious institutions in Nantes, from the College Saint-Stanislas to the Externat of the Enfants Nantais to the Pension des Dames de Chavagne. Had there really been "two different classes of youths" in a state of latent hostility as proclaimed by a slogan popular at that time? Even if I try hard to focus my thoughts on that matter, I am not at all sure. My fellow students and I had no qualms about thinking of ourselves as an aristocracy of the secondary school system; we scrutinized from afar, with some malevolence but without knowing or having any contact with its members, an elite we judged to be not as brilliant as our group (although wealthier), because it was taught by a faculty less endowed with diplomas and official titles, and thus, judging by their scholarly results, a faculty less reliable. The other state institutions—the Ecole professionnelle Livet, the écoles primaires supérieures, the écoles normales—were judged to be beneath ours, and quite frankly despised. All things considered, I believe we were aware that it was only a matter of time before our customary insults hurled against the "boîtes à curés," schools run by religious orders, would be a thing of the past, and that the two separate, parallel paths would eventually converge somewhere down the road. This rivalry—taking into account its irritable temper as well as its total inoffensiveness—brought to mind the ancient reciprocal allergy between the *noblesse de robe* and the *noblesse d'épée*, the religious and military hierarchies of the Old Régime, which had generated for the most part only a skin-deep rash. For me, the only sign of division was the pink-ribboned cap worn by the pupils of Saint-Stanislas on their school day outings (only lycée students wore complete uniforms: black, with golden palm leaves embroidered on the lapels), indicative of a separation so insignificant that adults left it up to their children for comment. As time passes, I cannot help asking myself if this image of a widely accepted social *consensus*, of an almost unanimous adherence to the established order has not stayed deep inside of me in the form of a hidden, fervent yearning. Just like humanity's belief in a fabulous golden age, there remains inside of me, secreted away in the crypts of memory, the image of an original, pacifist, social Eden from which some flaming sword has ousted me since adulthood, a paradise lost forever. It seems odd that I acquired this image during my contact with a Manichean city which has, for the last two centuries, participated in national uprisings to a point of exasperation, a characteristic that seems to be its mark of distinction.

I was well aware that within that vast class of people with means—modest means—whose children attended the lycées and secondary schools (a prerogative that limited, and also exactly defined it)—there existed a privileged sector distinguished by the possession of influence, prestige, money, and land, and by the power it exercised in decisions concerning fashion and social mores, but I could only vaguely identify it. The lycée was not a good field of observation to arrive at some clear ideas concerning this matter; besides, in Nantes itself, this sector of society was divided into two segments which had almost no contact with each other. Every time I travel down the coast starting at the south of the Loire's delta, from Pornic to the point of Saint-Gildas and especially along Préfailles, I am struck by how the coast has been marked here by the original character of its summertime residents' properties. There are no beaches except for a few small, sandy stretches hardly attractive for bathers, only a long line of small, irregular cliffs a few meters above the rocky ground covering punctuated by bulky rocks, cliffs cutting into thickets of brush and shrubbery which quickly turns reddish brown under the summer sun, a coastal area where narrow roads twist up- and downhill between tall hedges, stands of heather, and fern groves. Old-fashioned, rather large estates extend up to edges of the low cliffs, surrounded by walls built right up to the dropline, and accessible only from inland. The houses are hidden from view, jealously tucked away behind the bowers and shrubbery of their small parks: dramas played out on the open sea can only be seen from a theater of spacious, rigidly compartmentalized private boxes whose major concern seems to be how to conceal all movements and activities of their occupants from the view of neighbors. It would probably not be wrong to guess that the people summering here, residents whose private lives are so discreetly walled in, are not interested in cocktail parties or five o'clock teas, yachts, luxury cars, or trips to play baccarat at the casino, that they prefer to spend long hours of leisure time quietly under the trees, and that their distractions are limited to visits from neighbors and acquaintances of the same ilk and persuasion, people with whom ties have been established a long time ago through hereditary friendships and close family relations. No newcomers or occasional summer visitors spend their vacations here on the Jade Coast, only *families* (Préfailles is *the family beach*); they are the long-time owners of summer residences with unrestricted views of the sea, as well as the proprietors of impressive crypts from which to contemplate afterlife, in either the Bouteillerie or Miséricorde Cemetery. These old families from Nantes or Brittany who live on inherited wealth will no doubt continue to discreetly haunt the coast of Préfailles, even if their means are no longer what they once were; a little-known aristocracy which kept to itself, made no waves, and was hardly visible—not even visible in the lycée's classrooms: its children only made occasional appearances there after the baccalaureate, to attend advanced math classes required before entering the Ecole Navale or Ecole Polytechnique, courses not offered at the religious schools they attended because those schools lacked qualified faculty. I remember reading their names on the first pages of the Commencement Exercise program announcing the prize winners—but only on the pages listing names of the graduating class—the flourish of ancient, unusual names seldom encountered in a *bastion* of secularism: Urvoy de Portzamparc or Kersauzon de Penendreff. Besides, students in special preparatory classes for the national entrance exams who also boarded were members of a privileged caste, already half at liberty since they were authorized to study during recreation periods, and allowed to go out on Sundays without being accompanied by a *designated person*. We could only catch glimpses of these distinguished fellows in the hallways—remote, discreetly polite, already sporting that halo which separated those destined to enter a Grande Ecole from the rest of us—or watch them through the windows of their *study hall*, standing on one leg in front of one of the blackboards which covered the walls like armor, wearing white, ink-stained coats, chalk in one hand and eraser cloth in the other, lost in deep meditation over an algebraic rebus.

Except for the estates which adorn the banks of the Erdre, Nantes, as already mentioned, does not have any pleasant suburbs dotted with summer residences, nothing that resembles the elegant villas and baroque castles visible from the train going west from Angers. The Loire's wide, gray delta with its flat embankments, completely Nordic in its aspects, is unsuitable for vacation purposes. However, the powerful attraction of the beaches, which started at the end of the last century, has separated the residential suburbs from the city and pushed them farther west, like Le Pyla in Bordeaux's Bay of Arcachon. And, just like those villages in Corsica built with a view of the coast which have an *outpost* on the seashore that carries the same name, I have never been able to think of Nantes without including what I consider its annex: that seaside fringe of beaches reoccupied by the city at the start of each summer—a mental image so convincing that the stretch of territory separating them strikes me as a dead zone, one of those urban voids created temporarily during periods of demolition and renovation. However, the physiognomy of the city's beachfront which stretches from Le Croisic to La Bernerie changes considerably from the north to the south of the estuary. The somnolent south coast, without casinos, without dance halls, without places of entertainment, repopulated in the summer with little fanfare by an old, established bourgeoisie whose wealth consists primarily of landholdings, families who summered there every year in their familiar surroundings, keeping up traditional relationships and long-established habits, was quite the opposite of the north coast where, from Pouliguen to Pornichet, everything was in perpetual motion for at least two months. There were parties, special events, distractions, and a good deal of vanities on display by a summer population who lived along the crescent-shaped embankment of the beach at La Baule, and who provided a spectacle of its own, like the public of a sold-out theater. Something I could not quite distinguish or identify while living in Nantes because the equalizing hustle and bustle of a big city prevented it from taking shape revealed itself, concentrated itself here in the summer; it was as if a display window had been set up for two months in the vicinity of the casino, between the avenue de la Plage and the avenue des Lilas, in which a roster of Nantes' elegant people had taken up their posts, complete with labels stating the cost of their summer holiday.

At that time, the embankment had not yet become that roaring river which has since transformed the seashore along most of the French beaches—shores and cliffs no longer beaten by waves, but flooded by cars. It almost resembled a pedestrian zone, where beach dresses, the white trousers of tennis players, tussor shirts, and colorfully striped *blazers* lent it a festive air, reminiscent of the Balbec of *Les Jeunes Filles en fleur*, so different from the ubiquitous nudism of the nineteen-eighties. That elegant, more than affluent Nantes society making its ritual appearance in La Baule during the summer (almost like a publicity appearance) was represented by department store owners, shipbuilders, captains of industry, bankers, officials of the press, and luminaries from the world of entertainment—all of them attracted by and seeking the limelight just as avidly as the old money avoided it. Although I had not the least contact with that society, and looked at it only from afar, it was less strange for me than the other because quite a few of its sons attended the lycée as day students. I don't think I was more prone to a naïve snobbishness than the one usually affecting children between the ages of twelve and eighteen, something which is apt to cause frictions at home; but I have always had a tendency to look spontaneously for examples of *inimitable lives* among big baked goods factories or large foundry empires rather than among the landed gentry; Nantes gave me a taste of milieus described in Morand's *Lewis and Irene* which I discovered at that time. All things told, these stars of the social merry-go-round, whom I did not seek out and could not identify with certainty, did not provoke any particular hostility on my part just because we remained strangers. Paris is full of social faux-pas and blackouts, and utterly vulnerable to treacherous, hidden schemes like the one narrated in *Histoire des Treize*; but it also remains the dramatic battleground where a Rastignac, after his soliloquy in the Père Lachaise cemetery, chooses to wage war. I left the provinces carrying within me the image of a society characterized by a rather peaceful, not rigid stratification, where only the first step really counted: that initial step up from the social void into a sphere where the layers overlapped and still remained distinctive, layers that stayed in contact with each other and which, here and there, were quite amenable to infiltration. That *social upper crust* was unknown to me, but did not seem at all inaccessible because I had already come in contact with it, though only under formal circumstances; I looked at it as an outsider, without envy, without familiarity, and without hostility. For me, the least comprehensible fact in *Le Rouge et le Noir*, which I read with passionate interest during those years in Nantes, was the hero's drive to succeed at all cost in the world, and his unhealthy determination to rise in social circles. In 1925, it looked to me as if the defense mechanisms of Nantes' high society no longer necessitated breaking down any barriers in order to gain access, if one's mind was really set on it.

# I HAVE ALWAYS PAID

very close attention to the progressive changes in the landscape which announce the approach of a city. Especially when traveling by train, I am on the lookout for those first signs of infiltration, eager to see how the city's feelers stretch probingly into the countryside; and, if it happens to be a city where I like to live, I look upon them as a hand raised in welcome, waving from afar on friendly soil. Every year, on the way to Pornichet where we spent our summer vacation, I would catch a first glimpse of the town—just the crowns of a few isolated pine trees rising above tall hedges—while the train was still traveling through a dreary inland landscape; next came a few freshly painted fences, and then suddenly three or four villas, shockingly white against the trees, like Arab houses in a desert palm grove (today, it is quite difficult to imagine how French villages used to hide their silhouettes in those times; they barely made an impression on the eye before melting into a uniformly colored patch of gray like a flotilla camouflaged during the war). Although the moment I stepped off the train I immediately found myself in a more invigorating, livelier, and more festive world, in the midst of a well-dressed, sun-tanned, local crowd in wraparound skirts, beach dresses, and bright saris, the first, modest sign of arrival had been the view from the train, that welcoming wave of a hand raised for just a moment. Years later, arriving in New York from inland, nothing struck me more than the sight of a valley lying half under water, still in a state of almost total wilderness without a single house in sight, which the train crossed immediately south of the Palisades, less than two kilometers from the Pennsylvania train station located right in the heart of the city.

It is possible, even probable, that with the passage of time my interest in geography made me pay greater attention to how a city is anchored to its country-side, and to the complexity of exchanges it maintains with the milieu from which it originally drew its substance. Nantes, not my hometown, was first of all a city where I came to live for certain periods of time and then left again, every arrival and departure accompanied by feelings of violent alienation, liberation, or anguish. The whistles of the locomotives at the central train station, audible day and night from everywhere in the lycée, never let me forget it. Even when I lived there, the city remained more of a horizon than a milieu; to leave, and especially to return, was like a brief and stormy embarkation on a private odyssey: I could only think of Nantes as a city which, unlike any other, *announced* itself to me from afar in a powerful and sometimes oppressive way.

Because of the way it is situated, Nantes' relationship with its hinterland, as well as with the surrounding countryside, is completely different from that of its sister-cities located at the other two deltas of the Atlantic Ocean, Rouen and Bordeaux. Every time I see Rouen from high up on the road along the coast of Sainte Catherine, I am struck by the unique character of its site because it looks like an emblematic short-cut of the entire Haute Normandy region. Every major feature of Normandy's landscape is represented here, lined up on adjoining spaces: the river—the cliffs of chalk already visible at La Roche-Guyon and even at Meulan, rising on opposite sides of the river like a system of visual echos—the great forests down in the plains surrounded by the wide loops traced by the river—the star burst–like ravines of the little valleys once covered by cotton fields and dotted with red brick factories like beads on a string, valleys that reach deep into the suburbs to establish a foothold, like vines trying to put down roots or the trails of organic suckers, shoring up the city and draining toward it the riches of the plateau. In Bordeaux, even if a first glance at its site does not strike the visitor as being particularly impressive, the economic ties of the city to its rural environment are even more evident because of the storage facilities at the port, the warehouses of the great vineyards like the "châteaux" of the Médoc and the Pavé des Chartrons. Rouen, like Bordeaux, is an authentic regional metropolis that stands for and personifies its surrounding province; both are centers of worldwide exchanges, and almost a source of inspiration for a certain rhythm and lifestyle carefully cultivated and continuously renewed. Mauriac between Malagar and Bordeaux, just like Flaubert between Croisset, Yonville, and Rouen, could well be cited as examples illustrating the well-regulated traffic and smooth dynamics which here unite the country and the city into an all-inclusive, representative whole, like the owner's residence to his estate.

Nantes' relationship with its hinterland is entirely different. History represents a first stumbling block. Rouen, Caen, and Normandy were solidly behind la Gironde in 1793, just like Bordeaux and the rest of its *département*—whereas the city of the Loire unanimously took up arms against its rebellious rural population. In May 1968, the spontaneous, reciprocal hostility arose once again, as strong as ever: surveillance posts were put in place at the entrance points to the city in order to monitor entry and exit movements: a historical reflex, and an age-old reminiscence. These are also signs of a radical incompatibility of disposition, caused by the very poor representation of the centrally located regulatory bodies in the provinces. Though once the capital of the former dukes, Nantes had been deeply humiliated by Rennes at the time of the monarchy, which looked down on it from atop its Parliament and its *Etats de Bretagne*; around 1920 it was no longer the seat of any truly prestigious institutions of provincial power: there was no academy, no appellate court, no archbishopric. Only the restructuring of the university and the creation of regional *préfectures* have, very recently, lifted it to some degree out of that *deminutio capitis*.

This disgrace was caused largely by natural forces. There is no unity in the adjoining regions which Nantes could influence; ignorance and even open mistrust permeate the constitutive elements of their mosaic. The metropolis, very recently elevated to capital of the "pays de Loire," does not rule its river. Starting at Le Cellier and La Varenne, fifteen kilometers upstream, the Loire becomes a lifeless canyon traveled only five or six times a week by a solitary tanker carrying fuel up to Angers; it looks like a narrow valley kept under surveillance by a row of "high and ancient villages" looking down from the summit of their hills, settlements Vidal de la Blache considered to be "troublesome like a barrier." During the war of the Vendée, the river here carried gunboats, but traffic has never been completely restored since then; it already languished during the last century. Downstream, all along its estuary, components of Nantes' contingent of factories are scattered into clusters which continue to proliferate anarchically: it is an industrial, feebly condensed nebula still infiltrated to a large extent by the countryside. Although they face each other from opposite sides of the city, the Loire of the oil refineries and the Loire of the eel fishermen turn their backs on each other, ignore each other: traveling from Saint-Florent to the beach at Pornichet at the beginning of summer vacation, I was always surprised to see the juxtaposition of the estuary's building sites and factories with the old rural burgs which kept to themselves, clustered around their obese village churches like women gathering their skirts against the splashes of puddles. The huge mountains of dirt, the trenches and gaping craters dug by bulldozers which today have become an integral part of life in the countryside, were a most unusual sight then (a time when everyone was still unaccustomed to changes), reminiscent of a settlers' revolt. The short trip took me from a sort of rather dark, shaded valley reminiscent of those surrounding Tours, where willows and riverbanks succeed each other in a bucolic parade, to a delta resembling those of the North Sea, under a low sky, and fouled by yellow and gray smoke. The Loire does not really have a proper mouth; it resembles a flat, flared Baltic fjord kept under lock and key upstream by Nantes' bridges and landfills, a stream cautiously colonized by industry which grows in spurts, and moves ahead by maneuvering around the old villages through the remaining free spaces.

Going in the other direction, from north of the Loire to the south, even though one remains in the land of limestone and farmlands criss-crossed by hedges and trees, one has the impression of entering another country: the sudden passage from slate to tile roofs is accompanied by a less spectacular change in *gradation*, but which completely alters within a distance of a few kilometers not only the use and allocation of the land, but also the choice of crops, the looks of the houses in the villages, and even their occupants' lifestyle and way of thinking. I have already spoken of the wine country on the other side of Saint-Sébastien where vineyards contemplate each other across the river Sèvre, of its villages awash with sunlight between cool, shady areas underneath their trellises, tiled roofs, and fig trees, a region where one is immediately attuned to the native ambiance of a genuine, naïve *joie de vivre*; long ago, Charette's *Paydraits*, his soldiers recruited from the Pays de Retz beyond Lake Grandlieu, had already scandalized the devout population in Angers and the plain of Mauges by their bawdiness and fondness for dancing and merrymaking. The Loire-Atlantique's southern region, along the banks of the Sèvre or the Vallet and Mouzillon vineyards, is for Nantes—whose spires and towers can be glimpsed from here—a first, timid sign on the horizon of France's south, the *Midi*; a vision briefly evoked also by the Italian atmosphere of the mirror-like waters, Roman tiles, brickwork, and arcades in the small town of Clisson. But up north, right after crossing the river, everything changes. It is rare to find an open country as incurably gray and bleak, and as determined to wallow in the silence of a deep, rural sleep as those areas stretching from Nantes to Redon, from Nantes to Blain, from Nantes to Chateaubriant and to Segré: on that side, it seems as if an impenetrable barrier, an isolating *border* separates Nantes not only from the heart of Brittany, but even from the Gallo-Morbihan region, something which renders the often voiced claim for reattachment to the old Celtic province so inconvenient and problematic. Looking at those farmlands and hedges, at those dull, monotonous, lifeless pastures, a compacted, massive expanse of earth with no rivers, there is nothing left to do but observe, like René-Guy Cadou from his window in Louisfert, "the great rush of lands toward the end of the horizon"; a region where the town (coming from the north, one enters Louisfert almost unaware of having arrived) never was able to inject sparks of life into the surrounding countryside. It seems as if a *curfew* had been imposed during broad daylight on lands untilled and unplanted, forgotten by history and the economy; walking across it, one almost expects to see pigs rooting for acorns under the occasional, dreary stand of oak trees. But this is by no means an indigent province, far from it; its sleepy looks are deceiving. It has accomplished a tour de force by modernizing its economy without in the least altering its image of dreariness and pettiness. This is not one of France's zones of poverty, it is a zone of lifelessness: it seems as if no excitement could ever provoke a reaction here, a land where all echoes die out. The insurrection of 1793, which in a few days set the entire department south of the Loire on fire, came to a halt at the river's

189

firewall, with barely a few burning cinders flying across it, which were quickly extinguished on the north bank.

Among all those regions surrounding Nantes, regions which almost ignore her, where her influence is almost nil, and from which she is alienated by a noticeable incompatibility of disposition, there is not a single one she can call home; there are no traces of long-established links of dependence and assistance, of exchanges of mutual services, no tracks of people coming and going which tie a domain to its master's house. Aside from the floodplain at the lower Erdre, south of Nort, and the area around Lake Grandlieu, where one senses the city's proximity, where its spires and towers are clearly visible on the horizon from the marshes' flatlands, the *Pays Nantais* hardly exists. Aside from what is produced by the truck farms stretched around the beltways, the city draws very little from a region which is not without riches. Nantes drinks muscadet, the local wine, now raised to the rank of official national product, but never bothered much to promote it: cane sugar, cacao, rum from the Antilles, and bananas interested her far more. Surrounded by France's most impenetrable lands, regions which care not in the least about what happens elsewhere but also resisted exterior influences for the longest time, the city, for more than three hundred years, has found no compelling argument to form a harmonious whole with her provinces; and although her citizens, just like everywhere else, came from the neighboring regions, they have kept the least amount of mud on their shoes. I once compared Nantes to a grand port *culterreux*—a great hicktown-by-the-sea, a city tenaciously infiltrated by the countryside, as confirmed by her character which is much more plebian than aristocratic. As a matter of fact, throughout the nineteen-twenties, the drawn-out accent of the Vendée farmlands and the peasant-like slowness of gestures encountered in Nantes' inner suburbs more than once indicated a very recent transplantation. But there was no trace of the recent arrivals' former condition and mores: nothing to remind one of the downcast resignation of the peasant population who still lived almost like serfs out in the farming communities. As in a medieval town, "the air of the city"—here more than elsewhere—is known to "emancipate"; drawing from human stock in the surrounding communities, one could say that the city had risen, taken shape long ago to rebel against them, fight their traditional values and sedentary virtues. It is that implant in a great maritime and commercial

191

city surrounded by rural communities fast asleep, getting by on an agriculture of subsistence as would a city of Greece's Golden Age besieged by indigenous malevolence, which confers on Nantes the cutting autonomy, the air of boldness and independence which fills her streets, easy to perceive but difficult to define. Among the neighboring cities, Rennes and Angers have for too long a time slumbered in the opulence of landownership, and done little to build up or nourish such an ambiance; La Rochelle is not sufficiently populated. I have tried, by this rather long preamble, to describe the air of freedom, just like the breeze that fills the sails of a ship, which filled my lungs on the city's streets, and which I still breathe there today. When I lived there as a youth, I knew that my stay was only temporary and therefore had little desire to develop any attachments; but no other city was better suited to uproot a young life at an early age, to unfold and take apart the world's partitions before one's eyes: all the journeys imaginable—far beyond those described by Jules Verne—found their ideal point of departure in that adventurous city.

Finally, in my opinion, the lack of solid outside support has served Nantes well. Any attempt to link it to a territorial movement or provincial influences is an exercise in futility. As we have seen, Nantes is neither an integral part of Brittany, nor of the Vendée, and not even a city of the Loire, in spite of the artificial creation of the region labeled "Pays de Loire," because she blocks off, rather than vitalizes, an inanimate river. This is an advantage because it has made Nantes —together with Lyon (although Lyon is infinitely better integrated within the general circulation of its territories) and especially Strasbourg—one of the least provincial great cities of France. Deprived of any real osmosis with the neighboring countryside, freed from the narrow economic constraints of a local market, I used to think of her as *the city*, a city more than any other independent of her natural supports, camped like a stranger in her own territory, without the least care about how to get along with it.

Some time ago, after rereading one of Mauriac's early novels, *Préséances*, I realized to what extent Bordeaux had been for him the perfect setting for his life; an environment where, right from the beginning of his career, he found everything he needed. A self-sufficient setting which gently but firmly suggested to all those within earshot that it provided both the site and the modus operandi, that it lacked absolutely nothing of whatever was needed to assure and accomplish a life's

193

destiny. Leaving the city in order to look for fame and fortune elsewhere, such as the young Mauriac's move to Paris, could only be an act against nature, a gesture of rupture, of sheer violence against the local, unwritten law and order of things which made Bordeaux the one and only adequate and necessary living space for its natives. This is a primary example of the archetypal castrating, possessive mother complex—a city obsessed by jealousy, just like the heroine of Mauriac's *Génitrix*. Nantes was smaller, less populated, less monumental, not as strongly linked to her surrounding countryside, less ostensibly hierarchized in her society, and less grounded in her function as a provincial capital; a city which provided an immediate opening to the outside world since one did not have to pass through some intermediate territory. Thus the city's function for me was not so much maternal, but rather that of a matrix: after my seven years of statutory incubation, she set me free to seek a larger horizon, without breaking my heart, and without drama; it was a separation, an expulsion that left no scars. As a matter of fact, that separation did not take place at the end of my last school year in Nantes during the ritual brouhaha of the annual distribution of prizes, an event which then only seemed to lead once again mechanically, as in previous years, to the customary two and a half months of annual vacation. It happened the following year when, after a year spent in Paris, I took a trip to London where I was supposed to perfect my English during summer vacation. It was the first time I had ever left France, a departure I clearly remember—and recalled when I wrote the beginning of *Le Rivage des Syrtes*. Since I planned to take an early train, the express Bordeaux-Dieppe which stopped in Nantes at the station *Gare de l'Etat*, I had asked my great-aunt, who had been the *designated person* into whose care I was released on alternate Sundays during all those years at the lycée, if I could spend once more the night at her house. When I shut the garden gate of the rue Haute-Roche for the last time behind me, the day was just beginning; it had that limpid, serene, peaceful air of a morning after early prayer service, when the only sound one hears are birds singing in the trees, an ambiance the title of a novel by André Dhôtel (which I have not read) never fails to evoke for me: *Les rues dans l'aurore.* An empty tramway rumbled down the route de Rennes, sounding like the isolated buzz of a lone bee in flight. The emptiness of the streets in the early morning hours, a vision never seen before, seemed magic; walking in the city, in the calm, fresh morning air, felt like strolling the dew-covered alleys of a garden while everyone in the house was still asleep. After reaching the Morand Bridge, I turned into the quai des Tanneurs and then continued down the quai d'Orleans; when I crossed Feydeau Island, a yellow-pink ray of sunshine reached over the top of the houses' façades on my right. Street after street, the city took leave of me with a smile; the time had come to say good-bye, an *adieu* accompanied by a feeling of lightness without shadows. Our differences settled, we bid each other farewell, united in that carefree song of dawn. I had not been happy here, but neither did I feel that those years were something to be thrown overboard; I had absorbed, stored so much inside of me. My heart went out to these silent streets, those familiar, sinuous indentations inside that mold I was about to leave: it was not just a city where I had grown up, but a city where, at odds with and against her, according to her, but always with her, I had taken shape.

# HAVING LIVED IN A MUCH

too close symbiotic relationship with Nantes, years during which my mental image of the city became more and more detailed and extensive as I grew from childhood to adulthood, it is not surprising that I have difficulties arriving at a definitive picture. Rather, the city's image tends to define *me*: generally speaking, whatever the time frame, I never *see* myself other than completely immersed, body and soul plunged into a much more solid, more stimulating, and at the same time more restrictive element than what is usually referred to as the *milieu*. I try to distance myself from that intricate web of streets and places that had such a profound bearing on me when I was young and most impressionable: a risky, uncertain enterprise, because all that which closely touched our early formative years never completely ceases to participate somehow—be it ever so slightly, and even from afar—in our mutations.

Since I spent my early childhood in a rural environment, I am very much aware of differences in tension that separate the country from the city. The country is not just a sedative milieu (or was, at least until a few decades ago), characterized by the rarity, and, at the same time, the relative insignificance and placid character of the visual and sonorous signals it emits; it is basically a neutral field that tends to mold life into a vegetative form, and to impose habitual actions on social relationships. In the city, tension is caused by the concentration of living spaces, by the state of latent, continuous friction which galvanizes relationships, by the bewildering range of *possibilities*, of choices offered to the individual, and, most of all, at least for me, by the antagonism between a system of natural, centrifugal *outward drives* which channel the flow of the urban center's energy toward peripheral points of dispersion and the powerful centripetal force which counterbalances it, and thus maintains the city's cohesion. In such a concept of opposites, *The man of the crowds* dear to Edgar Poe could symbolize one of the poles: he incarnates the imbalance resulting from total submission to a central, urban attraction. On the other hand, there is no doubt that a centrifugal attraction regulates the powerful movements of escape and re-entry, reflected in today's travel mania and the exodus on holidays and weekends.

This is how the image of Nantes reconstructs itself dynamically in my memory—a little like a spider weaving its web: first the radials I so often traveled, starting from the center which represents the hub, the raw core of that double attraction; then the parallel lines of the side streets which fuse together and homogenize the ensemble, the connecting streets less often walked, the shortcuts, alleys, paths, and passages that dissect and capriciously cut up the urban mass, a network familiar to the true city dweller but which I never knew completely. Inside that web, place-names emerge in a disorderly fashion to reclaim their position and fasten themselves into place. They act like powerful beams of light, pulling a large area of the city out of the shadows like a piece torn from a map, showing itineraries walked so often they could never be forgotten, as well as unrelated snapshots, without a connecting theme; but instead of reassembling themselves into an image of the city, they project on it a canvas riddled with holes, covered by opaque zones that resemble badly developed, unidentifiable negatives.

First of all, the names, beginning with the name of the city. Too familiar for many years to be perceived in its singularity, but which, after being heard only here and there and less often, has once again taken on an aura of distance and independence which intrigues me; almost like a woman who, by reverting to her maiden name after a divorce, inadvertently reveals certain aspects of her personality and seems to regain her youth. A name more dense then resonant, with a great capacity to absorb and encompass because of the long, open *a* in its middle, which conveys a certain plenitude to the nasal syllable around which its articulation takes place. A name much more feminized by its inflection than I first thought, with a rather vague, floating outline, but which the connotation of plurality nuances with a discrete opulence, meaningful in its substance but hardly inclined to flaunt it. A name also feminized and saturated by water, and by the strong nautical connotation of its sonority, an impression reinforced long ago in my mind by the city's emblem painted on the creamy yellow tramways, the figure of a ship under sails with the motto *Favet Neptunus eunti*. A name much more closely tied to the liquid element than the city itself, and which more than any other (though unjustifiably) adds luster to many songs of the old nautical folklore. A city difficult to size up or pin down, wrapped up in her softly cushioned name as if protected by a layer of bubble wrap. Not entirely landlocked, and not strictly a seaport: neither fish nor fowl—just the right chemistry to create a mermaid.

★

There are few cities where successive municipalities have misused as shamelessly the privilege of naming and renaming places and avenues as they have in Nantes, and transformed at will a repertory of streets into obituaries of dead or rejected town councilors. The old names which escaped this misappropriation by elected officials have become that much more endearing, easy on the ear, and fondly remembered. Their frequent endings in *ière*, in *eau* or in *ais*, so well suited to the rural toponymy of the region (La Tortière, la Jonnelière, la Morrhonière—la Barbinais, La Refoulais, La Hunaudais—le Landreau, le Grand Blottereau, le Port Communeau) call to mind those enclaves in the midst of an urban expanse which were once part of the countryside, areas that have remained intact, and kept their original freshness. They remind me of a layer of native soil, from which the city secretly continues to draw its life blood. In the nineteen-twenties, that layer of topsoil was not just something that had barely been plowed under and still survived in the realm of place-names: my great-aunt still counted a number of *petits rentiers proprets* among her acquaintances, dapper elderly gentlemen with small independent incomes who wore celluloid collars and *cravates "à système"* (pre-knotted ties attached to shirt collars), who lived like she did in a small house with a garden, and received rents from their tenant farmers as well as *redevances*—payments in kind in the form of lumps of butter, chickens, and eggs by the dozen, produced on farms around Nozay, Sautron, or Carquefou. These old-fashioned names, drawn from deep within the soil, still touch me; they carry traces of an umbilical cord which the city, in spite of its zeal for emancipation (*Favet Neptunus eunti*), has never been able to cut completely.

Superimposed on that urban humus are the names of medieval Nantes, all of them gathered within the narrow perimeter bordered by the former quays (now filled in), the Château, the rue de la Marne, and what was once the riverbed of the Erdre; I have already mentioned a few of them on these pages: rue des Echevins, rue de la Juiverie, rue du Petit Bacchus. A purely archeological Nantes emerges here, a site with no echo, no harmonious ties to other parts of the city, and just as strange for Nantes' citizens as Lutèce can be for today's Parisians. For me, Nantes' most evocative names are those which, each in its own way, have broken their ties with historical or anecdotal origins in order to form a purely verbal constellation inside of me, where the city's layout finds itself imprisoned and exalted, like those ancient maps of the sky which superimpose the outlines of the Dog, the Bear, or Orion

202

the Hunter over the clusters of stars: Pirmil Bridge—
rue Kervégan—marché de la Petite Hollande—quai
de la Fosse—cours Saint-Pierre—Port Communeau—
Morand Bridge—quai d'Orléans—place Royale—passage Pommeraye—rue Crébillon—rue du Calvaire—
place Graslin—marché de Talensac—rue Félibien—
Sainte-Anne—Saint-Similien—Saint-Nicolas—Saint-
Clément—place Bretagne—place Viarme—rue du
Marchix—rue Monselet—rue des Dervallières—place
Canclaux. It is without doubt the toponymy, arranged
like a litany, and the sound of the place-names suc-
ceeding each other which enables memory to proceed
with the linking process, and draw a picture inside of
us that corresponds to our *idea* of a city we no longer
inhabit. My vision of that idea is a compact agglomeration, narrowly and badly dissected by streets, never
at rest and filled with sounds (bells ringing on Sundays
when we went on our outings), a compact urban block
inhabited by a bourgeois population, touched only
here and there by industry and the sea traffic, and infiltrated by placid little suburban gardens as soon as one
starts to leave the inner city. A city that has much
more in common with a Dutch town, with its bourgeoisie and their tulips, than with a city in Spain, since
private enterprises and the many-layered, industrious
buzzing of a thousand domestic beehives here cover

up and almost drown out the arid *Dienst* of its official services. There is nothing in the confusing, democratic jumble of its edifices to remind us of cities whose crenellated walls immediately call to mind their austerity, asceticism, and hierarchy—church, army barracks, convent, citadel, hospital, prison—all those centers of cloistered activity grouped together which rise in front of the visitor in Lérida, Pamplona, or Segovia. Much to my surprise, I found an example of this kind of cityscape not long ago in Langres, where an austere, monumental cliff crowns part of the old city sitting high up on the hill, right above the track of its former funicular railroad. Nantes is an aggregate of a strictly civilian society, a vivacious, disorderly, unplanned proliferation, born of its own spontaneous vitality, without an official decree of state like Le Havre or Cherbourg, and without the discipline of a merchants' confederation like the Hanseatic towns. A city which invented herself, which reinvents herself every day, without being particularly anchored in her past, and without being excessively fixated on her memories. Having forgotten, in due time, the stopover of Julius Caesar (now the name of a café), the sieges of Alain *Barbe-Torte*, the stakes of the Middle Ages and its Jewish ghettos; forgotten also, the islands and commerce of ebony wood, the wars of the Vendée, and now forgetting its railways and its transporter bridge and even its river, which today flows alongside the city at a distance, like a rejected suitor. Faithful only to a blind vitality, to the force of inertia within its masses moving forward, which continue to propel her ahead. A city without rough edges, without anything protruding from its compact mass (except for the exotic incongruity of the Tour de Bretagne), where even the cathedral has neither towers nor spire; a city which has come a long way, and yet a city where the local dialect and signs of peasant awkwardness can suddenly resurface at a street corner, just as simply and spontaneously as an overnight crop of mushrooms. Perhaps nothing but a substance, a living, nourishing substance to be plunged into again, something like an ever active, fermenting bread dough: a city forever renewing herself. Valéry Larbaud, who briefly visited Nantes in the nineteen-twenties—during the time I lived there—only remembered (aside from the passage de la Pommeraye) what the city has since done away with: "Nantes has an immense river, divided into several arms by islands that are covered by houses and streets as far as the eye can see, though they are just the city's suburbs. There is also a remarkable glass-enclosed passage, a passage several floors high, theatrical, with iron staircases that connect floors and provide access to boutiques with beautiful, shining windows, arranged around the galleries like glass cases in a museum. Finally, the trains pass along a quai that runs through the middle of the city, alongside the streets; they all seem to be express trains about to join the ocean liners ready to set out to sea. This is like the America of the novels of Jules Verne (who was born in Nantes)—America before and after the war of Secession—the America of long, pointed beards and caps with short square visors, of dark blue uniforms with facings and braiding in white worn by the infantry, yellow for the cavalry, and red for the artillery—an extraordinarily modern America, and which will always remain modern thanks to Jules Verne—but it would be preferable if the locomotives passing through the streets of Nantes had snowplows and big bells."

Less suggestive than the garland of intertwined, emblematic place-names draped around a city, powerful names whose evocation bring it back to life (just like the often quoted children's rhyme in the old song "Orleans—Beaugency—Notre Dame de Cléry—Vendome!—Vendome!" capable of resuscitating the fifteenth century and the Loire of the Book of Hours) are the snapshots which appear on that screen inside of us whenever we try to project pictures of a familiar town. The spontaneous ways in which memory sorts and groups them is a reminder that a sequence of intelligible sounds is far more likely to result in sheer poetry than sights mechanically registered by the eye and rigidly framed by one's perception. A stack of postcards spread out *flat* in front of us, even "personalized" postcards, could never succeed in reconstituting the overwhelming, indivisible, sheer mass of the city as seen in our mind's eye; it would only dehumanize, devitalize that vision. Oddly, my interiorized *views* of Nantes have not been updated: they refuse to take into account the transformations which took place in the city over the last fifty years; rather than ordinary memories, they constitute an archival ensemble of intimate documents, classified and inventoried. The first of those images is no doubt a view of the Erdre flowing into one of the Loire's former eddies, seen from the narrow quai d'Orleans, almost on the axis of the little river, against a background of the old town houses on Feydeau Island. It is a typical "Tableau parisien" à la Baudelaire, nothing but surfaces of stone and a body of calm water, without the various touches of greenery which were added after the landfill. Next comes the former view of the Loire as seen from the place de la Duchesse Anne, at the foot of the Château, looking toward the tip of Feydeau Island: almost like a smaller version of the *Monnaie* in Paris as seen from

the edge of the Ile de la Cité. This snapshot prominently features the sloping embankment of the old
quai Baco across from the Château, weeds crowding
the paving stones shaded by trees in the summertime.
The third view is from the current café on the place
Royale, an image of the houses on both sides of the rue
Crébillon, including, beyond the place Graslin, the
crenellated space of sky cut out by the façade of buildings at the entry of the rue Voltaire; in that privately
held *negative*, the place Royale has kept its big clock
and its configuration before the bombings, when it
was slightly smaller than now. I can almost hear the
lively talk and laughter of the Sunday crowd assembled there in the summer—even noisier than today's
crowds because of the chirruping on the large terrace
of another café at the corner of the rue La Pérouse, the
café d'Orléans, which has since disappeared. Neutral
images from which life has withdrawn, but where the
blocks of buildings, the bodies of water, the fixed
arrangement of empty and occupied space frozen in
time stand out conspicuously, looking almost dis-
dainfully at the crowds milling about. They effec-
tively rejoin the ghost-like, mineral qualities of *le
Rêve Parisien*—an insidious x-ray taken by visual
memory, where only the skeleton remains after all
the adjoining tissues have been eliminated. Images

personalized by our presence, but which neverthe-less remain nonchalantly vague and noncommittal about man's impact on his physical surroundings, they evoke the haughty, majestic neutrality of empty streets during the early morning hours.

The truth is that neither the magic of names nor the snapshots engraved in my memory will ever allow me to embrace and recapture the city in her entirety. I have lived there wrapped up inside my imagination rather than surrounded by reality: her physical pres-ence has been for me what a first garrison might be for a second lieutenant who dreams of being chief mil-itary commander someday. It was an environment filled with premonitions, where everything was a sign or a symbol and each barrier an obstacle to overcome, a place where *to be alive* was first and foremost a pow-erful yearning to grow up, *to develop.* A city that has watched over your fledgling debuts will fade and re-cede into the background unless memory has been able to capture what happened during the time it pro-vided the unique, irreplaceable, familiar warmth of an incubator. A broken egg can be put together again, a cocoon with holes can be mended; but nothing will ever be able to re-create that blind urge inside which willed everything around me to burst or explode so that I could learn to exist in other ways, nothing will ever be able to re-create the plasticity, the malleability of a soul still vague, on which every impression became an imprint, or rather, in a Goethean sense, a printed *form*, a destiny in the process of developing itself.

Perhaps it would be better to allow Nantes to reshape herself inside of me simply by calling on chance encounters with various bits of flotsam and jetsam, sometimes imaginary, sometimes real; all of them the result, or fallout, of the same explosion. They would come together, helter-skelter, without any kind of organization: the dream-like vision of the omnibus in the film *Zéro de Conduite*, which takes the student back to his boarding school. The odor of cold coal and dense winter fogs settling at dusk on a city illuminated by pale strings of streetlights. The ceramic, modern-style décor which continues to transform *La Cigale* into a provincial Lipp, reduced to the size of a candy box. The round paving stones, the cloistered little houses of the old passage Russeil, more silent than a nuns' convent behind its garden gates, underneath its magnolia trees. The winding, hairpin-shaped rue de *Garennes* which dominates from up high where the Loire flows past Trentemoult, and its counterpart in my memory, the panorama of St. Augustine in Florida, as illustrated in the Hetzel edition of Jules Verne's novel *North against South*. The small covered tramway stop on the place du Commerce, where the yellow cars went screeching around the bend in the tracks with a noise that made one's teeth grit—a stop that served as nighttime shelter for a bum who wrote poetry, author of at least one unforgettable line:

Greetings to you, roses that bloom in the snow!

The beginning of the rue Charles Monselet, and its connecting elbow to the boulevard Delorme, a spot which announces from afar the peace and quiet of the area around Procé Park, where the frantic activities of the inner city subside as suddenly as if one had momentarily entered a secluded garden: a site of glad tidings, the promise of a pleasant neighborhood—and the exact opposite of the boulevard de Doulon, which has always impressed me as the harbinger of gloom. The poster announcing the lyrical program for the week, fastened behind its protective screen to the wall of the Graslin theater's colonnade, almost identical in its format and lettering to the straw-yellow posters of the Comédie Française. Although nothing seems to link these disparate, at times ridiculous images which bear not the slightest resemblance to each other, I think of them as an old gold piece cut into quarters and then into smaller and smaller wedges: as if the city, after bursting into a myriad of pieces, had reconstituted herself into an entity more meaningful than what could ever be achieved by any and all panoramic views remembered. The secret of this process lies in the sovereign, instinctive manner in which a young, impressionable soul still without a guide, without models, and without being urged, sorted out and made its selections from the chaos at hand. Not a single one of the city's emblematic images has been able to anchor a landmark inside of me by connecting it to a specific date in my past, because I developed no close ties, formed no personal attachments, nothing except an almost abstract drive of the *ego* to annex and absorb, an enormous, acquisitive, prospective bulimia which dominates life from age eleven to eighteen. I was growing, and so was the city all around me, engaged in the process of changing and remodeling herself, carving out her boundaries, setting her perspectives, forging ahead in a manner amenable to any and all future expansions—the only way for her to remain inside of me, and still be herself—and to keep on changing.

212